Remote Sensing Image Fusion

A Practical Guide

Remote Sensing Image Fusion

A Practical Guide

Christine Pohl
John van Genderen

CRC Press
Taylor & Francis Group
Boca Raton London New York

CRC Press is an imprint of the
Taylor & Francis Group, an **informa** business

CRC Press
Taylor & Francis Group
6000 Broken Sound Parkway NW, Suite 300
Boca Raton, FL 33487-2742

First issued in paperback 2019

ISBN-13: 978-1-4987-3002-0 (hbk)
ISBN-13: 978-0-367-87361-5 (pbk)

Library of Congress Cataloging-in-Publication Data

Names: Pohl, Christine, 1966- author. | Van Genderen, J. L. (John L.) author.
Title: Remote sensing image fusion : a practical guide / Christine Pohl and
John Van Genderen.
Description: Boca Raton, FL : Taylor & Francis, 2016. | Includes
bibliographical references and index.
Identifiers: LCCN 2016013861 | ISBN 9781498730020
Subjects: LCSH: Remote sensing--Handbooks, manuals, etc. | Image
processing--Handbooks, manuals, etc.
Classification: LCC G70.4 .P65 2016 | DDC 621.36/78028566--dc23
LC record available at https://lccn.loc.gov/2016013861

Visit the Taylor & Francis Web site at
http://www.taylorandfrancis.com

and the CRC Press Web site at
http://www.crcpress.com

Contents

Foreword

When Christine Pohl asked me to write a foreword to this book, I realized that almost 30 years had gone by since I was involved in my first publication on image fusion. Roy Welch and I published a paper on pansharpening of Landsat Thematic Mapper data using a SPOT panchromatic 10-m resolution image. We actually did not call it pansharpening or fusion but "merging."[1] Because none of the image processing systems of that time offered an algorithm for image fusion, I had to program an intensity-hue-saturation transform to be able to create a merged 10-m resolution Landsat-4 TM/SPOT-1 image of Atlanta, Georgia. About 10 years later in 1998, a larger number of pansharpening methods existed, prompting Christine Pohl and John Van Genderen to publish the then state-of-the-art review paper on image fusion, which is probably the most cited paper in this field.[2] They brought structure and principles into the fusion efforts by an ever-growing number of authors. It also took some time before the remote sensing community realized that fusion techniques are not restricted to its field.

In the best object-oriented terminology, "fusion" is a polymorphic term. To quote Edwards and Jeansoulin's foreword to a special issue of the *International Journal of Geographical Information Science*: "Data fusion exists in different forms in different scientific communities. Hence, for example, the term is used by the image community to embrace the problem of sensor fusion, where images from different sensors are combined. The term is also used by the database community for parts of the interoperability problem. The logic community uses the term for knowledge fusion."[3] Other fields embracing the term "fusion" include medicine, military science, automobile, and aviation industry, just to name a few.

It is therefore good and fitting that Christine and John limit the fusion discussion to its application in remote sensing, coining the term "remote sensing image fusion" (RSIF) as the topic of their new review publication. This time they decided that the breadth is too great to cover RSIF in a journal paper and went for a textbook instead. It is exciting to see that after 20 years, their basic structure for fusion levels is still valid, although fusion papers and methods have mushroomed since 1998. I recommend this book to scientists, students, and teachers who want to inform themselves about developments and the state of the art in image fusion at a structured and understandable level. The authors provide not only a thorough and in-depth analysis of more than 40 different fusion algorithms and methods including the equations but also show the importance of appropriate preprocessing techniques—an issue that is often overlooked in image fusion discussions.

A full chapter is dedicated to quality assessment—a topic as hotly contested as the development of the "best" fusion algorithm. The authors show that often the quality indicator that shows the best results for the respective author's fusion technique is used. It is, however, important to know that there is no single best fusion algorithm as there is today no best fusion quality indicator. Quality is always related to the application. Consequently, Christine Pohl and John Van Genderen devote a chapter on application with a detailed analysis of case studies from different remote sensing fields. It is evident that this field has moved on from simple pansharpening to a general RSIF discipline including radar, LiDAR, and hyperspectral data sets. Fusion algorithms have moved from simple one-formula techniques to complex hybrid and multilevel approaches.

I congratulate Christine and John on accomplishing the exhaustive task that they undertook in putting this book together. We now have the new definitive state-of-the-art textbook on remote sensing image fusion. Twenty years from now, it will probably serve as the new reference point from where to start the next scientific progress report. Christine Pohl and John Van Genderen have brought order into the R&D of image fusion.

Manfred Ehlers
Professor emeritus for GIS and Remote Sensing Chair
Competence Center for Geoinformatics in Northern Germany
Osnabrück, Germany

References

1. Welch, R. and M. Ehlers, 1987. Merging multiresolution SPOT HRV and Landsat TM data. *Photogrammetric Engineering and Remote Sensing* 53 (3):301–303.
2. Pohl, C. and J. Van Genderen, 1998. Review article. Multisensor image fusion in remote sensing: Concepts, methods and applications. *International Journal of Remote Sensing* 19 (5):823–854.
3. Edwards, G. and R. Jeansoulin, 2004. Data fusion—From a logic perspective with a view to implementation, Guest Editorial. *International Journal of Geographical Information Science* 18 (4):303–307.

Preface

Remote sensing delivers multimodal and multi-temporal data from Earth's surface. In order to cope with these multidimensional data sources and to make the most out of them, image fusion is a valuable tool. It has developed over the past few decades into a usable image processing technique for extracting information of higher quality and reliability. As more sensors and advanced image fusion techniques have become available, researchers have conducted a vast amount of successful studies using image fusion. However, the definition of an appropriate workflow prior to processing the imagery requires knowledge in all related fields—that is, remote sensing, image fusion, and the desired image exploitation processing. From the results, it can be seen that the choice of the appropriate technique, as well as the fine-tuning of the individual parameters of this technique, is crucial. There is still a lack of strategic guidelines due to the complexity and variability of data selection, processing techniques, and applications. This book gives an introduction to remote sensing image fusion (RSIF) providing an overview of the sensors and applications. It describes data selection, application requirements, and the choice of a suitable image fusion technique. It comprises a diverse selection of successful image fusion cases that are relevant to other users and other areas of interest around the world. From these cases, common guidelines, which are valuable contributions to further applications and developments, have been derived. The availability of these guidelines will help to identify bottlenecks, further develop image fusion techniques, make the best use of existing multimodal images, and provide new insights into Earth's processes. The outcome is an RSIF textbook in which successful image fusion cases are displayed and described. They are embedded in common findings, and lead to generally valid statements in the field of image fusion. The book helps newcomers to obtain a quick start into the practical value and benefits of multisensor image fusion. Experts will find this book useful to obtain an overview on the state of the art and understand current constraints that need to be solved in future research efforts.

The book is relevant to newcomers and specialists alike. Students, scientists from other fields, and teachers will find a practical guide to enter this exciting technology. The overview, categorization, and definitions provided in Chapters 3 through 5 and the practical examples contained in Chapter 6 provide newcomers with a quick start. Experts and scientists in the field will benefit from the summary of latest findings and be guided to new research areas. Particularly, sections on "Selection Approach," "Communalities," and "Contradictions" in Chapters 4 and 5 will appeal to their interests. For industry professionals, the book can be a great introduction and basis for understanding multisensory remote sensing image exploitation and the development of commercialized image fusion software from a practical perspective. Owing to the wide coverage of Earth observation applications in a multisensory context, the book will be relevant to many different disciplines, for example, mapping, agriculture, forestry, urban planning, hazard monitoring, geology, mineral exploitation, and others. This book contains achievements in RSIF during the past decades. It explains the concept and mathematical background and adds a discussion on quality assessment of RSIF. Quality and operational issues of image and data fusion attract increasing attention as the integration of multimodal data has become a requirement and remote sensing data has entered day-to-day applications and form a major information provider. The outlook in Chapter 7 explains and discusses current trends and developments in image and data fusion. Topics, such as data mining, Big Data, and cloud computing, as solutions to handling the ever-increasing number and amount of data sources to derive meaningful information, are portrayed at the end. The handling of multimodal Earth observation data requires broad, interdisciplinary knowledge, teamwork, and data sharing and needs further research in all accompanying aspects. This means that data fusion remains an exciting field with lots of potential to further evolve and improve our understanding of the environment in which we live and to make wiser decisions in the future to live a sustainable life.

Along with this book we provide additional online material containing an extensive list of organized literature on the subject. Furthermore, there are presentation slides for each chapter to be used by teachers and lecturers who would like to use the book as textbook for their students. The website where the online material is located is *rsif.website*. In addition, it contains updates on ongoing discussions and research in the field of remote sensing image fusion.

Acknowledgments

This book is the result of numerous research projects, meetings, and conferences attended by the authors. The compilation of data from these diverse sources has made it possible to identify the needs, issues, techniques, and useful remote sensing image fusion applications and trends over the last 25 years during which the subject evolved. In addition, we would like to explicitly acknowledge a special group of contributors. These are the scientists who took the time to respond to an online questionnaire collecting information on their experience and common practices in remote sensing image fusion. We thank Gintautas Palubinskas, Christian Berger, Rajiv Singh, Saygin Abdikan, Shridhar Jawak, Flora Paganelli, Vladimir Petrović, Bo Huang, and Hadi Fadaei. Six of these investigators also provided material for illustrative case studies found in Chapter 6 on applications. We also gratefully acknowledge the work of Damdinsuren Amarsaikhan (fusion examples for urban application and forestry), Saygin Abdikan (complementarity study of optical and radar remote sensing for urban and agricultural areas in Turkey), Shridhar Jawak (pansharpening studies for San Francisco, California, USA and Larsemann Hills, Antarctica), Flora Paganelli (structural mapping with multi-angle radar imagery in Buffalo Head Hills in north-central Alberta, Canada), and Hadi Fadaei (tropical forest deforestation monitoring in Sarawak, Malaysia). Special thanks go to Gintautas Palubinskas who provided his expertise and a large contribution to Chapter 5 on quality assessment, describing a general framework in Section 5.5. Zhang Jixian provided the ZY-3 fused example illustrated in Chapter 7. All contributions have been clearly marked in the text with relevant references that can be consulted for further reading. We also acknowledge the efforts of Patrick van der Heiden for his contribution converting some of our sketches into illustrative figures. We thank Nor Nisha Nadhira Nazirun for her assistance in drawing the technical flowcharts in Chapter 4.

Authors

Christine Pohl obtained her BSc in geodesy at the Technical University of Braunschweig and continued her MSc at Leibniz University Hannover in Germany. She earned her PhD in remote sensing from an international program at the International Institute for Aerospace Survey and Earth Sciences (ITC), the Netherlands, together with Leibniz University, Hannover, Germany, in 1996. Between 1996 and 1998, she worked as deputy head of the research division at the European Union Satellite Center (EUSC) in Madrid, Spain, followed by a period as an associate professor at ITC. From 2002 to 2012, she was as CEO for a private company in Europe, where she was responsible for the educational program and provided coaching and leadership training for managers.

Currently, she represents the chair for remote sensing at the University of Osnabrueck, Germany. Her research interests cover mainly standardization procedures for remote sensing image and data fusion. She uses remote sensing for applications in the tropics, monitoring oil palm plantations, mapping, change detection, humanitarian crisis support, and coastal zone management. She serves on the technical committees of several international conferences in the field of remote sensing. Dr. Pohl is an editor for the *International Journal of Remote Sensing* and is a member of the editorial board of the *International Journal of Image and Data Fusion*.

John L. van Genderen graduated from the University of Queensland, Brisbane, Australia, after which he studied remote sensing at the ITC in the Netherlands for his MSc degree and earned his PhD in remote sensing from the University of Sheffield in the United Kingdom in 1972. Since then, he has conducted research, teaching, projects, and consulting assignments in more than 140 countries around the world.

He was a professor of remote sensing at the ITC from 1991 until his retirement in 2009. His main research interests in terms of technology have been in RSIF, image processing, and SAR interferometry, and in terms of applications in coastal zones and natural and man-made hazards.

Dr. van Genderen serves on many editorial boards of remote sensing journals, such as the *International Journal of Image and Data Fusion*, and has been active in many national and international remote sensing societies.

Acronyms

ALOS	advanced land observation satellite
ARSIS	Amélioration de la Résolution Spatiale par Injection de Structures (English: Improving spatial resolution by structure injection)
ASTER	advanced spaceborne thermal emission and reflection radiometer
AVHRR	advanced very high resolution radiometer
AVIRIS	airborne visible infrared imaging spectrometer
AWT	additive wavelet transform
BGP	band generation process
BT	Brovey transform
CALIPSO	cloud-aerosol LiDAR and infrared pathfinder satellite observations
CBD	context-based decision
CCRS	Canadian Centre of Remote Sensing
CD	change detection
CE	cross entropy
CMSC	composite based on means, standard deviations, and correlation coefficients
CNES	Centre National d'Études Spatiales
CNR	National Research Council
CORR	correlation
CS	component substitution
CTA	classification tree analysis
CVA	change vector analysis
DI	disturbance index
DoD	date of disturbance
DEM	digital elevation model
DLR	National Aeronautics and Space Research Centre
DMSP	Defence Meteorological Satellite Program
DSM	digital surface model
DSS	decision support system
DST	Dempster–Shafer theory
DTM	digital terrain model
DWT	discrete wavelet transform
ERGAS	Erreur Relative Globale Adimensionnelle de Synthèse (English: relative dimensionless global error of synthesis)
ESA	European Space Agency
ESTARFM	enhanced spatial and temporal adaptive reflectance fusion model

ETM+	Enhanced Thematic Mapper Plus
FCC	false color composite
FCNFS	Fisher criterion-based nearest feature space
FFT	fast Fourier transform
FPCS	feature-oriented principal component selection
GIHS	generalized IHS
GLCM	gray-level co-occurrence matrix
GLI	Global Imager
GP	Gaussian pyramid
GPR	ground penetrating radar
GPS	global positioning system
GS	Gram–Schmidt fusion
GSD	ground sampling distance
H	entropy
HCS	hyperspherical color space
HPF	high-pass filtering
HPM	high-pass modulation
HR	high resolution
HRG	high-resolution geometric
HRPI	high-resolution panchromatic image
HRS	hyperspectral remote sensing
HSR	high spatial resolution
HSeR	high spectral resolution
HSI	hyperspectral image
IAAS	infrastructure as a service
ICA	independent component analysis
IHS	intensity hue saturation
k-NN	k-nearest neighbor
LCA	linear combination approximation
LEDAPS	Landsat Ecosystem Disturbance Adaptive Processing System
LiDAR	light detection and ranging
LP	Laplacian pyramid
LR	low resolution
LRMI	low-resolution multispectral image
LSR	low spatial resolution
LST	land surface temperature
LULC	land use land cover
MERIS	medium-resolution imaging spectrometer
MI	mutual information
MIF	multi-sensor image fusion
MF	morphological filter
ML	maximum likelihood
MLC	maximum likelihood classifier
MODIS	moderate-resolution imaging spectroradiometer
MRA	multiresolution analysis

MSAVI	modified soil-adjusted vegetation index
MTF	modulation transfer function
NASA	National Aeronautics and Space Administration
NDLI	normalized difference landmass index
NDSI	normalized difference snow index
NDVI	normalized difference vegetation index
NDWI	normalized difference water index
NIR	near infrared
NSCT	nonsubsampled contourlet transform
NOAA	National Oceanic and Atmospheric Administration
OBIA	object-based image analysis
P+XS	pansharpening
PAAS	platform as a service
PALSAR	Japanese phased array L-band synthetic aperture radar
PC	principal component
PCA	principal component analysis
PCC	percentage of correct classification
PCS	principal component substitution
PSNR	peak signal-to-noise ratio
QI	quality index
RCM	Radarsat Constellation Mission
RE	ratio enhancement
RF	random forest
RF	regression fusion
RGB	red green blue
RMSE	root mean square error
RSC	relative spectral contribution
RSIF	remote sensing image fusion
RVS	regression variable substitution
SAAS	software as a service
SAR	synthetic aperture radar
SD	standard deviation
SFA	spatial filter approximation
SFIM	smoothing filter-based intensity modulation
SIFT	scale invariant feature transform
SIR	spectral index ratio
SMOS	soil moisture and ocean salinity
SPM	subpixel mapping
SST	sea surface temperature
STAARCH	spatial and temporal adaptive algorithm for mapping reflectance change
STARFM	spatial and temporal adaptive reflectance fusion model
STDFA	spatial temporal data fusion approach
STRS	spectral–temporal response surface
SVM	support vector machine

SVR   synthetic variable ratio
SWIR   shortwave infrared
UDWT   undecimated discrete wavelet transform
UIQI   unique image quality index
UNB   University of New Brunswick Fusion
VIR   visible infrared
VNIR   visible near infrared
WT   wavelet transform
WT–PCA   wavelet–principal component analysis (hybrid fusion)

1

Introduction

This chapter introduces the different concepts in remote sensing, with emphasis on the differences and complementarity of multiple sensors. Furthermore, remote sensing image fusion is defined in the context of data fusion. The chapter proceeds to provide a definition of remote sensing image fusion and describes the types of imagery commonly fused. This chapter also explains the purpose and objective of image fusion, and lists some of the main benefits and limitations of remote sensing image fusion. The main factors that need to be taken into account when fusing different types of remote sensing imagery are also introduced. A special role is given to active and passive remote sensing but also other types of sensors, such as thermal, hyperspectral, and light detection and ranging, are explained.

1.1 Outline of the Book

This book deals with remote sensing image and data fusion. It is assumed that the reader is familiar with remote sensing principles, techniques, and applications. However, as many readers of this book may come from other disciplines such as medical image processing, computer graphics, security, and defence, in Section 1.2, we give a brief introduction to remote sensing multi-sensor data. For more details on remote sensing principles, techniques, and application, the textbooks listed in Table 1.1 serve as a good introduction to this field.

In this introductory chapter, we introduce the topic of remote sensing image and data fusion, providing information on the objectives of image fusion, followed by a description of the types of remote sensing image fusion available today, plus discussions on the limitations of such fusion approaches. Also, the benefits and the many applications of remote sensing image fusion methods are briefly described. The chapter ends with some definitions and terminology used in this book and provides a list of literature references.

Chapter 2 discusses the various levels of remote sensing image fusion. After a brief introduction, the chapter explains and describes the various levels at which images and data may be fused. These include pixel level, feature level, advanced decision level fusion, as well as a section on data fusion.

TABLE 1.1

Textbooks on Remote Sensing

Title	Reference	Publisher	ISBN
Remote Sensing Handbook	Thenkabail (2015)	CRC Press	9781482218015
Remote Sensing and Image Interpretation	Lillesand et al. (2015)	Wiley	9781118343289
The Core of GIScience: A Systems-Based Approach	Dopheide et al. (2013)	ITC	9789036537193
Introduction to Remote Sensing	Campbell and Wynne (2011)	Guilford Press	9781609181765
Fundamentals of Satellite Remote Sensing	Chuvieco and Huete (2009)	CRC Press	9780415310840
Principles of Remote Sensing	Tempfli et al. (2009)	ITC	9789061641837
Remote Sensing—Models and Methods for Image Processing	Schowengerdt (2007)	Academic Press	9780123694072

Again, as for Chapter 1 and for all subsequent chapters, this chapter ends with a summary and a list of relevant literature.

In Chapter 3, the authors deal with the many preprocessing steps required prior to fusing different data sets. Key aspects treated in this chapter include the issues involved in selecting the appropriate data sources to be fused, the sensor-specific corrections that need to be made, such as geometric changes, and explanation of the several image enhancement techniques commonly used in remote sensing image fusion.

One of the key chapters of this book is Chapter 4 on the actual remote sensing image fusion techniques. This chapter commences with a novel categorization of image fusion techniques. As there are hundreds of image fusion algorithms that have been developed over the past 25 years, we have grouped them into several logical categories and then describe in detail each fusion algorithm in each of these categories. The categories described include (a) component substitution, (b) numerical methods, (c) statistical image fusion, (d) modulation-based techniques, (e) multiresolution approaches (MRA), (f) hybrid techniques, and (g) others. In all cases, we have tried to provide the original algorithm for the described fusion technique. After this detailed analysis of the many fusion techniques used today, the chapter concludes with a section giving guidelines on the selection approach and discusses the communalities and contradictions of the remote sensing image fusion algorithms.

Before going on to deal with the many applications of remote sensing image and data fusion, Chapter 5 deals with the important topic of quality assessment. It explains the various image quality parameters used to evaluate image fusion products and discusses and presents the main existing, established indices. A section is also devoted to the requirements and procedures for visual, subjective evaluation of fused image, and data products.

Chapter 6 is another major chapter in which we present the many applications of remote sensing image and data fusion. This chapter provides numerous case studies, showing the features, benefits, and results of fusing different types of data sets for an improved interpretation of the area under consideration. As for all chapters, this chapter also ends with some conclusions about the actual and potential applications of image fusion, and provides many references for the reader to consult for more details about any of the applications considered in this chapter.

The final chapter in this book gives an insight into the future. It presents some of the main trends and developments in remote sensing image and data fusion and introduces some new key technologies, which will influence this field over the coming years. This includes topics such as the remaining challenges and trends, data mining, cloud computing, Big Data analytics, and The Internet of Things.

Further reading is possible using the references provided at the end of each chapter. The references refer to the topics discussed in the individual chapter. The background for the information provided in this book is threefold. The foundation is built from the authors' many years of expertise in practical use of remote sensing image fusion (RSIF). A second information source is an extensive and dedicated database of journal and conference papers built over the years covering published material until 2015. This database enabled the categorization of RSIF as a research field, splitting the subject into various domains of interest, that is, techniques, sensors, applications, areas of achievement, and journal/conference. The latter led to an overview of which journals published the most RSIF papers and helps the reader to further deepen the subject of interest (see Figure 1.1). The purpose of RSIF and its research focus are subjects of Section 1.4.3. Other interesting information retrieved from the database with an overview of most popular techniques and sensors will be displayed in Chapter 4. Most common applications are discussed in Chapter 6.

1.2 Remote Sensing

"Remote sensing is the art, science, and technology of observing an object, scene, or phenomenon ... without" actually being in "physical contact with the object of interest" (Tempfli et al. 2009). The sensors are designed to provide information on a received signal. The signal depends on the materials on the ground with their unique molecular composition and shape. Electromagnetic radiation is reflected, absorbed and emitted differently, depending on the surface composition. Different sensors deliver different "views" of Earth's surface. The difference is given by spatial, spectral, and temporal resolution, view angle, polarization, wavelength, interaction with

FIGURE 1.1
Distribution of RSIF literature in international peer-reviewed journals.

the objects, and atmospheric influence on the signal. The advantage of using spaceborne remote sensing is the ability to acquire data over large areas, providing a synoptic view of Earth, in a multi-temporal fashion, allowing change detection methods to model complex Earth processes. Remote sensing distinguishes between passive, for example, visible and near infrared (VIR), or thermal infrared (TIR), and active, for example, synthetic aperture radar (SAR) and light detection and ranging (LiDAR) sensors. Active sensors emit their own waves while passive sensors collect emitted radiation and reflections of illuminated surfaces. The energy source for the latter is the sun. The wavelengths, bands, and frequencies of the different remote sensing systems are displayed in Figure 1.2.

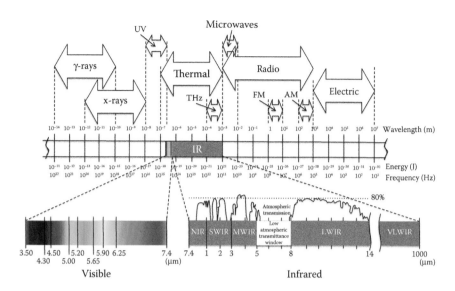

FIGURE 1.2
Electromagnetic spectrum with wavelengths and frequencies used in remote sensing.

1.3 Multi-Sensor Data

Owing to the increase in operational satellites providing a multitude of sensors, RSIF has gained increasing importance over the years. It provides the proper framework to deal with the diversity of data and ensure an optimum information outcome. VIR, on one hand, provides multi-band observations influenced by sun illumination and cloud cover but similar to the human perception of objects. SAR, on the other hand, delivers texture, geometry, and moisture-sensitive information, rather difficult to interpret for untrained users. Especially in applications where the disparate nature of this data combination provides valuable input, the fusion of VIR/SAR can be worthwhile. To illustrate this, Figure 1.3 compares the images of optical and radar sensors on the same area. The examples show two sites: one over flat terrain in the Netherlands (Ameland); the other covers very mountainous terrain in Indonesia (Bengkulu). The two images of Ameland show the North Sea, coastline, agricultural fields, and an urban area. Especially, the water and the urban area show the scattering behavior of microwaves, in their sensitivity to roughness (sea) and the corner reflector effect (bright urban area). The Système d'Observation du Terre (SPOT) PAN on the other hand allows a detailed view on the agricultural fields, infrastructure, and housing. The Indonesian SPOT multispectral image depicts

(a)   (b)

(c)   (d)

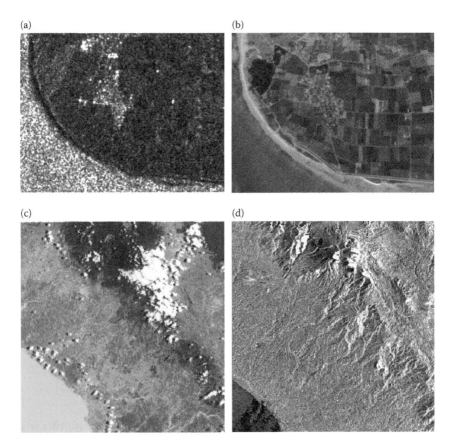

FIGURE 1.3
VIR and SAR images to illustrate the disparate nature of information acquired by passive and active remote sensing. (a) ERS-1 SAR and (b) SPOT PAN images covering the west end of Ameland, the Netherlands; (c) SPOT XS and (d) ERS-1 SAR covering an area near Bengkulu, Sumatra, Indonesia; SPOT imagery courtesy of SPOT IMAGE; ERS-1 SAR images courtesy of ESA.

the rich vegetation of the tropics (dark red), water (turquoise), and of course the cloud cover. Next to it, the ERS-1 SAR image provides a three-dimensional (3-D) perspective due to its texture and bright backscatter from the slopes that are facing the sensor. Later in Chapter 6, we will see the benefit of combining these images.

But also combinations of hyperspectral, thermal, and LiDAR data with other remote sensing images gain importance (Gomez-Chova et al. 2015). The open access to operational satellite sensor data, such as the National Aeronautics and Space Administration's (NASA's) Landsat and the European Space Agency's (ESA's) Sentinel, highly contributes to advances in multimodal image exploitation. The following sections will provide the necessary background to understand the individual sensors before we introduce RSIF.

1.3.1 Optical Remote Sensing

Remote sensing with visible and near-infrared electromagnetic waves takes advantage of the natural radiation of the sun that is reflected from Earth's surface. The sum of the reflected radiation, emissions from the ground, and the path radiance is sensed through the optics of the sensor and converted into electrical signals by detectors. Figure 1.4 shows the principle of optical remote sensing. These signals are collected, stored, and transmitted to ground receiving stations where the data are further processed.

As an example, Landsat Thematic Mapper (TM) data are acquired in seven different channels in the visible, near-, mid-, and thermal infrared spectra; the SPOT sensors 1–3 operated with three channels in the green, red, and near-infrared regions in multispectral mode, and in the visible spectrum in panchromatic mode. The wavelengths and frequencies are displayed in Figure 1.2. Common optical sensors with their characteristics, such as bands, wavelengths, spatial resolution, platforms, and revisit times, are listed in Table 1.2. The resulting VIR images allow the identification of ground cover types according to their so-called spectral signature. The detectability of objects in satellite imagery depends on the ground resolution and, in relation to this, the image contrast, which influences the representation of details of objects. Along with the spatial resolution, the spectral resolution of the data contributes to object identification and image interpretability. The characteristics of the area observed and seasonally dependent climate conditions play an important role too. Vegetation and land cover can be derived with high accuracy.

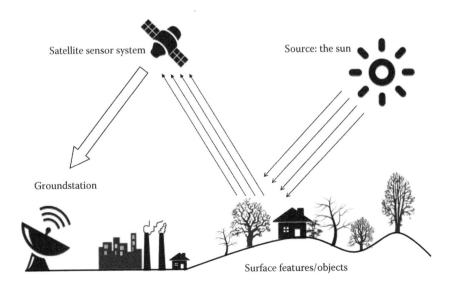

FIGURE 1.4
Principle of optical remote sensing.

TABLE 1.2

Popular Optical Remote Sensing Platforms and Their Specifications

Platform	Launch Date	Sensor	Bands	Wavelength (μm)	Spatial Resolution (m)	Revisit Time (days)
United States						
NOAA	1998	AVHRR	1	0.58–0.68	1090	2× daily
			2	0.725–1.00		
			3A	1.58–1.64		
			3B	3.55–3.93		
			4	10.30–11.30		
			5	11.50–12.50		
Landsat-8	2013	OLI	1 Ultra blue	0.43–0.45		16
			2 Blue	0.45–0.51		
			3 Green	0.53–0.59		
			4 Red	0.64–0.67	30	
			5 NIR	0.85–0.88		
			6 SWIR1	1.57–1.65		
			7 SWIR2	2.11–2.29		
			8 PAN	0.50–0.68	15	
			9 Cirrus	1.36–1.38	30	
			10 TIRS1	10.60–11.19	100	
			11 TIRS2	11.50–12.51		
Landsat-7	1999	ETM +	1	0.45–0.52		16
			2	0.52–0.60		
			3	0.63–0.69	30	
			4	0.77–0.90		
			5	1.55–1.75		
			6	10.40–12.50	60	
			7	2.09–2.35	30	
			8	0.52–0.90	15	
Landsat-4/5	1982/1984	TM	1	0.45–0.52		16
			2	0.52–0.60		
			3	0.63–0.69	30	
			4	0.76–0.90		
			5	1.55–1.75		
			6	10.40–12.50	120	
			7	2.08–2.35	30	
Landsat 1–3	1972 1975 1978	MSS	4	0.5–0.6	60	18
			5	0.6–0.7		
			6	0.7–0.8		
			7	0.8–1.1		
		PAN	1	4.50–9.00	0.6	
QuickBird-2	2001	MS	1 Blue	0.43–5.45	2.4	1–3.5
			2 Green	4.66–6.20		
			3 Red	5.90–7.10		
			4 NIR	7.15–9.18		

(Continued)

TABLE 1.2 (*Continued*)

Popular Optical Remote Sensing Platforms and Their Specifications

Platform	Launch Date	Sensor	Bands	Wavelength (μm)	Spatial Resolution (m)	Revisit Time (days)
		PAN	1	4.50–8.00	0.5	
			1 Coastal	4.00–5.10		
			2 Blue	4.50–5.10		
			3 Green	5.10–5.80		
WorldView-2	2009	MS	4 Yellow	5.85–6.25	2	1–3.7
			5 Red	6.30–6.90		
			6 Red edge	7.05–7.45		
			7 NIR1	7.70–8.95		
			8 NIR2	8.60–10.40		
		PAN	1	0.45–0.90	1	
			1 Blue	0.45–0.52		
IKONOS	1999	MS	2 Green	0.51–0.59	4	1
			3 Red	0.63–0.70		
			4 NIR	0.76–0.85		
		PAN	1	0.45–0.80	0.4	
			1 Blue	0.45–0.51		
GeoEye-1	2008	MS	2 Green	0.51–0.58	1.65	1–9
			3 Red	0.66–0.69		
			4 NIR	0.78–0.92		
			1	0.62–0.67		
			2	0.84–0.88		
			3	0.46–0.48		
			4	0.54–0.56	500	
			5	0.12–0.13		
			6	0.16–0.17		
			7	0.21–0.22		
			8	0.40–0.42		
			9	0.44–0.45		
Terra/Aqua	1999/2002	MODIS	10	0.48–0.50		0.5
			11	0.53–0.54		
			12	0.55–0.56		
			13	0.66–0.67		
			14	0.67–0.68	1000	
			15	0.74–0.75		
			16	0.86–0.88		
			17	0.89–0.92		
			18	0.93–0.94		
			19	0.92–0.96		
			20–36[a]	3.66–14.38		

(*Continued*)

TABLE 1.2 (*Continued*)

Popular Optical Remote Sensing Platforms and Their Specifications

Platform	Launch Date	Sensor	Bands	Wavelength (μm)	Spatial Resolution (m)	Revisit Time (days)
			1	0.52–0.60		
			2	0.63–0.69	15	
			3	0.76–0.86		
			4	1.60–1.70		
			5	2.14–2.18		
			6	2.18–2.22	30	
Terra	1999	ASTER	7	2.24–2.28		16
			8	2.30–2.36		
			9	2.36–2.43		
			10	8.12–8.48	390	
			11	8.48–8.82		
			12	8.92–9.28	90	
			13	10.25–10.95		
			14	10.95–11.65		
Europe						
			P	0.51–0.73	10	
	1986		B1 Green	0.50–0.59		
SPOT 1–3	1990	HRV	B2 Red	0.61–0.68	20	26
	1993		B3 NIR	0.79–0.89		
			PAN (Red)	0.61–0.68	10	
			B1 Green	0.50–0.59		
		HRV-IR	B2 Red	0.61–0.68	20	
			B3 NIR	0.79–0.89		
SPOT 4	1998		B4 SW-IR	1.58–1.75		2–3
			B0 Blue	0.43–0.47		
		VEGETATION	B2 Red	0.61–0.68	1000	
			B3 NIR	0.78–0.89		
			SWIR	1.58–1.75		
			PAN	0.48–0.71	5	
			B1 Green	0.50–0.59		
		HRV-IR	B2 Red	0.61–0.68	10	
			B3 NIR	0.78–0.89		
SPOT 5	2002		B4 Mid-IR	1.58–1.75	20	2–3
			B0 Blue	0.43–0.47		
		VEGETATION	B2 Red	0.61–0.68	1000	
			B3 NIR	0.78–0.89		
			SWIR	1.58–1.75		

(Continued)

TABLE 1.2 (*Continued*)

Popular Optical Remote Sensing Platforms and Their Specifications

Platform	Launch Date	Sensor	Bands	Wavelength (μm)	Spatial Resolution (m)	Revisit Time (days)
			1	0.41[b]		
			2	0.44		
			3	0.49		
			4	0.51		
			5	0.56		
			6	0.62		
			7	0.67		
ENVISAT-1	2002	MERIS	8	0.68	300/1200	3
			9	0.71		
			10	0.75		
			11	0.76		
			12	0.78		
			13	0.87		
			14	0.89		
			15	0.90		
			1 Blue	0.44–0.51		
			2 Green	0.52–0.59		
RapidEye (5 satellites)	2008	MS	3 Red	0.63–0.68	5	1
			4 Red edge	0.69–0.73		
			5 NIR	0.76–0.85		
			1	0.44[b]	60	
			2	0.49	10	
			3	0.56	10	
			4	0.66	10	
			5	0.70	20	
			6	0.74	20	
Sentinel-2a	2015[c]	MSI	7	0.78	20	10
			8	0.84	10	
			8a	0.86	20	
			9	0.94	60	
			10	1.38	60	
			11	1.61	20	
			12	2.19	20	
Japan[d]						
			1	0.42–0.50		
ALOS	1996	AVNIR-2	2	0.52–0.60	10	14
			3	0.61–0.69		
			4	0.76–0.89		

(Continued)

TABLE 1.2 (*Continued*)

Popular Optical Remote Sensing Platforms and Their Specifications

Platform	Launch Date	Sensor	Bands	Wavelength (μm)	Spatial Resolution (m)	Revisit Time (days)
China						
HJ1-1A	2008	WVC	1	0.43–0.52	30	4
			2	0.52–0.60		
			3	0.63–0.69		
			4	0.76–0.90		
		HSI	115[e]	0.45–0.95	100	
HJ-1B	2008	WVC	1	0.43–0.52	30	4
			2	0.52–0.60		
			3	0.63–0.69		
			4	0.76–0.90		
		IRMSS	1 NIR	0.75–1.10	150	
			2 SWIR1	1.55–1.75		
			3 SWIR2	3.50–3.90		
			4 TIR	10.5–12.5	300	
ZY-1 02C	2011	PAN	1	0.51–0.85	5	3
		MS	1	0.52–0.59	10	
			2	0.63–0.69		
			3	0.77–0.89		
ZY-03-01	2012	MSC	1	0.45–0.52	5.8	5
			2	0.52–0.59		
			3	0.63–0.69		
			4	0.77–0.89		
SJ-9A	2012	TDDICCD	1 PAN	0.45–0.89	2.5	4
			2 MS	0.45–0.52	10	
			3 MS	0.52–0.59		
			4 MS	0.63–0.69		
			5 MS	0.77–0.89		
SJ-9B	2012	IR Camera	IR	0.80–1.20	73	8
GF-1	2013	PAN	1	0.45–0.90	2	4
		MSC	2	0.45–0.52	8	
			3	0.52–0.59		
			4	0.63–0.69		
			5	0.77–0.89		
			6	0.45–0.52	16	2
			7	0.52–0.59		
			8	0.63–0.69		
			9	0.77–0.89		
GF-2	2014	PAN	1	0.45–0.90	1	5
		MSC	2	0.45–0.52	4	
			3	0.52–0.59		
			4	0.63–0.69		
			5	0.77–0.89		

(Continued)

TABLE 1.2 (*Continued*)

Popular Optical Remote Sensing Platforms and Their Specifications

Platform	Launch Date	Sensor	Bands	Wavelength (μm)	Spatial Resolution (m)	Revisit Time (days)
China/Brazil						
		CCD	1	0.45–0.52		
			2	0.52–0.59		
			3	0.63–0.69	20	
			4	0.77–0.89		3
			5	0.51–0.73		
CBERS-01/02	1999	WFI	6	0.63–0.69	258	
			7	0.77–0.89		
		IRMSS	8	0.50–0.90		
			9	1.55–1.75	78	26
			20	2.08–2.35		
			11	10.4–12.5	156	
		CCD	1	0.45–0.52		
			2	0.52–0.59		
			3	0.63–0.69	20	
CBERS-02B	2007		4	0.77–0.89		3
			5	0.51–0.73		
		HR	6	0.50–0.80	2.36	
		WFI	7	0.63–0.69	258	
			8	0.77–0.89		
		PAN/MSC	1	0.51–0.85	5	
			2	0.52–0.59		3
			3	0.63–0.69	10	
			4	0.77–0.89		
			5	0.45–0.52		
		MSC	6	0.52–0.59	20	
			7	0.63–0.69		
CBERS-04	2014		8	0.77–0.89		26
			9	0.50–0.90		
		IRMSC	10	1.55–1.75	40	
			11	2.08–2.35		
			12	10.4–12.5	80	
			13	0.45–0.52		
		WFC	14	0.52–0.59	73	3
			15	0.63–0.69		
			16	0.77–0.89		

(*Continued*)

TABLE 1.2 (*Continued*)

Popular Optical Remote Sensing Platforms and Their Specifications

Platform	Launch Date	Sensor	Bands	Wavelength (μm)	Spatial Resolution (m)	Revisit Time (days)
India						
		AWiFS	1	0.52–0.59	56	
			2	0.62–0.68		
			3	0.77–0.86		
			4	1.55–1.70		
Resourcesat-1/2 (IRS-P6)	2003/2011	LISS-III	1 Green	0.52–0.59	23.5	5–24
			2 Red	0.62–0.68		
			3 NIR	0.77–0.86		
			4 SWIR	1.55–1.70		
		LISS-IV	MS 2	0.52–0.59	5.8	
			MS3/Mono	0.62–0.68		
			MS 4	0.77–0.86		
Cartosat-1 (IRS-P5)	2005	FORE AFT	PAN[f]	0.50–0.85	2.5	5
Cartosat-2, 2A, 2B	2007, 2008, 2010	PAN	1	0.80–0.85	0.8	4

[a] Surface, atmospheric temperature, and cloud monitoring.
[b] Centre wavelength.
[c] Sentinel-2b planned for end of 2016.
[d] ASTER is a joint effort between NASA (USA) and JAXA (Japan)—see above.
[e] The detailed bands of the hyperspectral sensor are not publicly available.
[f] Detailed band descriptions are not available.

The essential constraint on optical remote sensing is the frequent occurrence of clouds, which prevents the sensors from collecting useful data. The low probability of acquiring nearly cloud-free SPOT or Landsat images limits the usefulness in timely needed information extraction. This is especially so in large parts of the world, such as the Tropics and the Temperate Zones. Only desert areas do not suffer this problem. In high latitude areas, the lack of daylight during the winter months is also a limiting factor in acquiring good quality optical imagery. The remote sensing user community therefore looked for alternatives to overcome this bottleneck. Radar data complement optical remote sensing are independent of weather conditions, and can acquire imagery both day time and night time. SAR imagery from space is now readily available, with plenty of different sensors to choose from. The information contributions of SAR sensors are discussed in the following section.

1.3.2 Radar Remote Sensing

The principle of SAR sensors is different to that of the optical remote sensing previously described. The radar sensor is an active sensor that emits

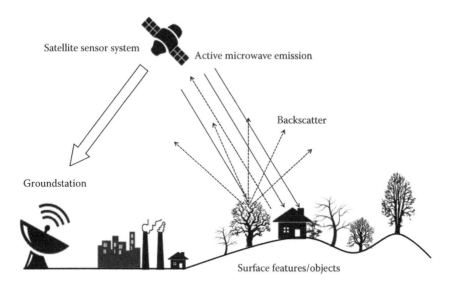

FIGURE 1.5
Principle of microwave remote sensing.

electromagnetic radiation with wavelengths of centimetres to meters using an antenna. The emitted energy passes through the atmosphere and reaches the ground, where it is scattered. Figure 1.5 illustrates the principle. Depending on the type of ground cover, more or less energy is backscattered in the direction of the radar antenna that receives the signal. The raw data are a combination of the power (amplitude/intensity) and phase of the backscattered signal whose arrangement depends on the slant range of the scattering object from the antenna. In fact, the measured element is the time which the signal needs to travel from the sensor to the object and back. From this information, the slant range between sensor and object can be derived using knowledge on the speed of light. The creation of the digital image is performed with the range (across track) and the time (in azimuth means along track) coordinates. The near range of an acquired scene is compressed with respect to the far range. The final pixel of a SAR image does not have the same dimensions as the resolution cell during data acquisition due to the variation of range resolution with incidence angle. Resampling is applied to form a uniform grid. The pixel spacing is not necessarily equivalent to the nominal resolution of the sensor.

Owing to the long wavelengths of active microwave sensors, SAR images contain information mainly on surface roughness, shape, orientation, and soil moisture. The image intensities depend on the illuminating signal characteristics such as wavelength, polarization, incidence angle, and scan direction, in addition to the properties of the illuminated surface (roughness, geometric shape, orientation toward antenna, and dielectric properties) (Dallemand et al. 1993). The various wavelengths are categorized in bands

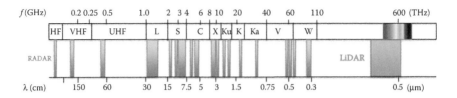

FIGURE 1.6
Radar bands in the electromagnetic spectrum.

as displayed in Figure 1.6. The letters that have been assigned to the various radar bands are from descriptions given in World War II but are still being used today.

Figure 1.7 helps to explain the tone information provided by SAR images. Smooth surfaces, for example, calm water bodies, appear black in the images (D stands for dark), while a strong wind can turn them into bright areas. The areas which were not illuminated by the microwave signal also appear black as shadow in the imagery. Medium tones (M) refer to a relatively rough surface, for example, forest canopy. Bright areas occur if more energy is scattered in the direction of the antenna caused by the so-called "corner reflectors" such as buildings (L for light). Radar image interpretation therefore requires a different approach from the conventional VIR data interpretation described above. Parameters like tone, texture, shape, structure, and size are all factors to be looked at (Dallemand et al. 1993).

The contents of SAR images vary as the system parameters vary. The role of the wavelength is illustrated in Figure 1.8. The longer the wavelength of the electromagnetic waves used, the larger the object they interact with. Hence, C-band (wavelength 3.75–7.5 cm) is more suitable for example in oceanographic applications (recognition of wave patterns, ice monitoring)

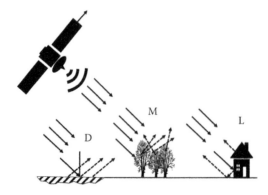

FIGURE 1.7
Interpretation of tone in SAR images—for explanations refer to the text. (Adapted from Dallemand, J. et al. 1993. Radar imagery: Theory and application. In *FAO Lecture Notes*. Rome, Italy.)

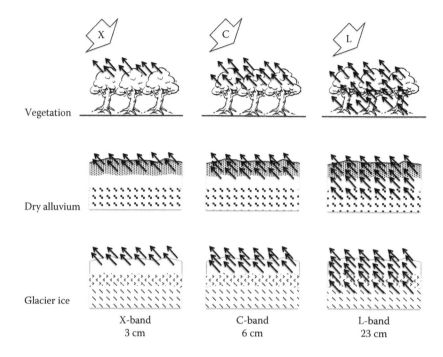

FIGURE 1.8
Wavelength influence on interaction of microwaves with natural targets. (Adapted from Dallemand, J. et al. 1993. Radar imagery: Theory and application. In *FAO Lecture Notes*. Rome, Italy.)

than L-band (wavelength 15–30 cm), which provides information on different forest stands for example interaction with branches rather than with the canopy. In addition, there is a difference in backscatter if a different polarization is used. In order to be able to judge the influence of polarization on the information content of SAR data, it is necessary to compare data that were acquired using similar system parameters but varying polarization.

Fully polarimetric SAR sensors provide four measurements in the linear horizontal (H) and vertical (V) polarizations HH, HV, VH, and VV (Touzi et al. 2004), as shown in Figure 1.9:

- HH—horizontal transmit and horizontal receive (a)
- VV—vertical transmit and vertical receive (b)
- HV—horizontal transmit and vertical receive (c)
- VH—vertical transmit and horizontal receive (c)

Polarimetric SAR information is directly related to physical properties of the observed scene and influences the backscattering mechanism (Du et al. 2015). Calibration of these measurements is necessary because they are affected by crosstalk due to the transmitting–receiving antenna or the switches used to

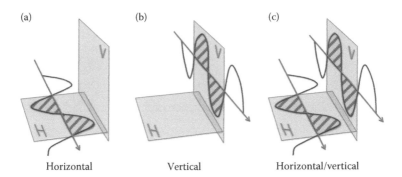

(a)   (b)   (c)

Horizontal   Vertical   Horizontal/vertical

FIGURE 1.9
Polarization of microwaves.

select the H or V polarization at transmission and reception (Touzi et al. 1993). The polarization signature can be used for image interpretation using polarization properties of points and distributed targets. The classification of fully polarimetric SAR images improves the quality of the results due to the different perspectives obtain on the surveyed area (Touzi et al. 2004).

Limitations of SAR are the sensitivity to mountainous terrain, that is, elevated objects. Due to the side-looking geometry, the backscatter from high elevations leads to foreshortening, layover, and shadows in the intensity image. The actual measured element in active microwave radar is the time the signal travels from the antenna to the target and return. The difference in time of two signals reflected from two different points of a slope (bottom and top) determines the size of the slope area mapped in the image. The distance between these two points (1 and 2 in Figure 1.10) is shorter for a steep

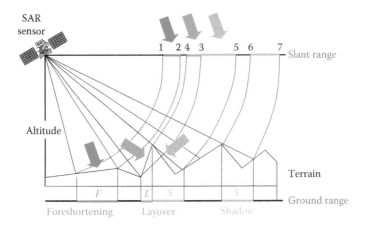

FIGURE 1.10
Layover, foreshortening, and shadow in SAR images. (Adapted from Schreier, G. 1993. *SAR Geocoding: Data and Systems*, Wichmann, Berlin.)

fore-slope than for a flat terrain. This effect is called foreshortening. It occurs in areas where the local incidence angle θ (Figure 1.10) is close to 90°. In terms of radiometry, foreshortening compresses the backscattered signal energy in the across-track direction (Schreier 1993). In extreme cases, the foreshortening turns into layover. Targets at the top of an elevated object have a shorter slant range with respect to the antenna than points lower down. Hence the signal backscattered from the top reaches the antenna before the signal coming from the bottom. The fore-slope is reversed (see points 3 and 4 in Figure 1.10). Radiometrically, these areas appear bright in the image because they face the antenna and image pixels are a combination of object backscatter superimposed. Radar shadow is different from shadow occurring in optical image data. Shadow appears where a back-slope is hidden from the radar signal, thus not illuminated. If the slope is steeper than the SAR incidence angle, the back-slope cannot be illuminated by the signal (points 6 and 7 in Figure 1.10). Shadow appears black in the image. Deviations from zero result from system noise. Obviously shadows in far range are longer than in near range. Under certain conditions layover can mask shadow zones, as shown in Figure 1.10 at points 3, 4, and 5. These geometric effects on the image radiometry have to be considered in image interpretation and processing. They certainly need to be taken into account when fusing such SAR imagery with other data.

Apart from the sensor viewing and surface geometry the amount of SAR backscatter is influenced by surface roughness and the dielectric constant of the surface material. Surface roughness is generated by the general topography and grain size of the surficial material. It is wavelength dependent if a surface appears rough in the image. The impact of surface roughness on the radar signal is illustrated in Figure 1.11. The dielectric constant depends on the moisture content and conductivity (Gaber et al. 2015).

Table 1.3 provides a list of radar bands with their specifications used in spaceborne remote sensing today, followed by Table 1.4 containing an overview on spaceborne SAR system wavelengths and frequencies related to their radar band.

1.3.3 Hyperspectral Remote Sensing

The term hyperspectral evolved with the availability of many small bandwidth (0.001–0.015 μm) image bands compared to a few bands of multispectral sensors with a larger bandwidth (0.05–0.012 μm), which is illustrated in Figure 1.12. Hyperspectral remote sensing (HRS), also called imaging spectrometry, imaging or reflectance spectroscopy, is very sensitive to narrow wavelengths and often collected in ten to hundreds of bands.

A definition for hyperspectral remote sensing is given by Goetz et al. (1985): "The acquisition of images in hundreds of contiguous registered spectral bands such that for each pixel a radiant spectrum can be derived." The idea is to convert the spectral information into physical properties based on reflectance and emissivity (Dor et al. 2013). The literature does not agree on

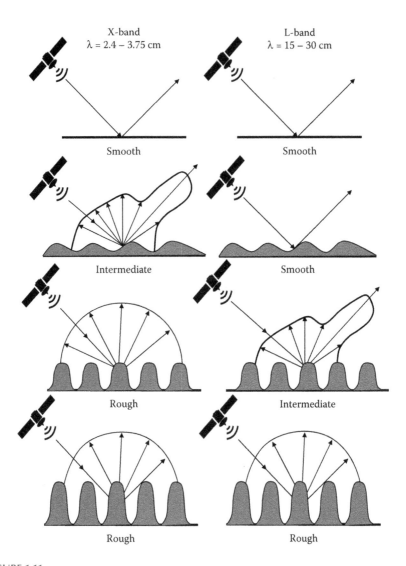

FIGURE 1.11
Influence of surface roughness on the backscatter of SAR.

TABLE 1.3

Common Radar Bands Used in Remote Sensing

Radar Band	Wavelength (cm)	Frequency (GHz)
X	2.5–3.75	8–12
C	3.75–7.5	4–8
S	11.11–7.69	2.7–3.9
L	15–30	1–2
P	100	0.3

TABLE 1.4

Spaceborne SAR System Wavelengths and Frequencies Related to Radar Band

Platform and Sensor	Operator	Band	λ (cm)	Frequency (GHz)	Polarization	GSD (m)
Sentinel-1 A/B 2014/2016	ESA	C	5.6	5.4	Dual	5
ALOS PALSAR 1/2 2006/2014	JAXA	L	23.62	1.3	Quad	10/3
Kompsat-5 2013	KARI	X	3.2	9.7	Quad	1
RISAT-1 ISRO C 5.6 5.35 Quad 1 COSMO-SkyMed 1-4 2007/2008/2010	ASI	X	3.1	9.6	Dual	1
TanDEM-X 2010	DLR	X	3.1	9.6	Quad	3
TerraSAR-X 2007	DLR	X	3.1	9.6	Quad	1
Radarsat-2 2007	CSA	C	5.6	5.3	Quad	3
Envisat ASAR 2002–2012	ESA	C	5.6	5.3	Quad	30
SRTM-C 2000	NASA	C	5.6	5.3	Dual	30
SRTM-X 2000	NASA	X	3.1	9.6	VV	25
Radarsat-1 1995–2013	CSA	C	5.6	5.3	Quad	10
JERS-1 1992–1998	JAXA (NASDA)	L	23.5	1.3	HH	18
ERS-1/2 1991–2000/ 1995–2011	ESA	C	5.6	5.3	VV	30
SEASAT 1978	NASA	L	23.5	1.3	HH	25

Note: Satellite SAR sensors that failed, are not yet launched, or have limited accessibility are not listed in this table.

the number of bands that distinguish multispectral from hyperspectral sensors. Often it is noted that 10 or more bands are called hyperspectral. Others mention more than 30 bands as distinction (Ciampalini et al. 2015). In HRS, the collected data are not represented by images but by so-called hyperspectral data cubes (see Figure 1.13).

Hyperspectral sensors have the advantage of providing contiguous bands in contrast to many multispectral sensors. Examples of hyperspectral sensors are listed in Table 1.5. Disadvantages are the large amount of bands to be interpreted with quite some redundancy, which requires particular processing

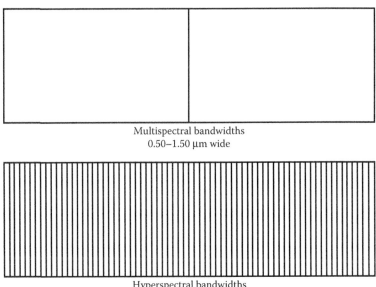

Multispectral bandwidths
0.50–1.50 μm wide

Hyperspectral bandwidths
0.01–0.15 μm wide

FIGURE 1.12
Small and large bandwidths of hyperspectral versus multispectral data.

methods. Algorithms are distinguished by four categories: (1) target/anomaly detection, (2) change detection, (3) classification, and (4) spectral unmixing (Manolakis et al. 2001). The reduction of dimensionality is performed in two ways: (1) band selection and (2) feature extraction (Kaewpijit et al. 2003). Band selection is discussed in Chapter 4. Dimensionality reduction

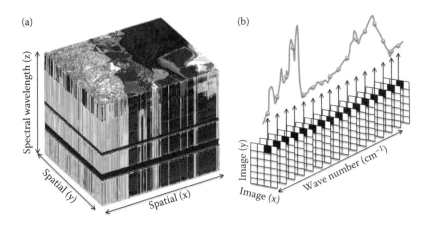

FIGURE 1.13
Hyperspectral data: (a) HSI data cube; (b) a three-dimensional image composed of spatial and spectral data. (Adapted from Bannon, D. 2009. *Nature Photonics*, 3 (11): 627–629.)

TABLE 1.5

Selected Hyperspectral Remote Sensing Systems

Operator	Sensor	Number of Bands	Wavelengths/ Spatial Resolution	Platform
NASA	Airborne Visible/Infrared Imaging Spectrometer (AVIRIS)	224	0.4–2.5 μm 4–20 m	Airborne
SPECIM	Airborne Imaging Spectrometer for Applications (AISA)	488 (Eagle) 254 (Hawk) 84 (Owl)	0.40–0.97 μm 0.97–2.50 μm 8–12 μm 1.5 m	
ITRES	Compact Airborne Spectrographic Imager (CASI)	288	0.42–0.96 μm <1–10 m	
GER	Geophysical Environmental Research Imaging Spectrometer (GERIS)	63	8–12 μm 15 m	
GER	Digital Airborne Imaging Spectrometer (DAIS)	72	0.4–2.5 μm 5–20 m	
Integrated Spectronics	Hyperspectral MAPper (HyMAP)	126	0.4–2.5 μm 5 m	
Selex ES	Sistema Iperspettrale Multisensoriale di Galileo Avionica (SIM-GA)	769	0.4–0.25 μm 0.167–1.5 m	
NASA	Hyperion (EO-1)	242	0.4–2.5 μm 30 m	Spaceborne
ESA	Compact High Resolution Imaging Spectrometer (CHRIS)	19	0.415–1.050 μm 20	

Source: A full list of imaging spectrometers is available from Kramer, H. J. 2012. *Observation of the Earth and Its Environment: Survey of Missions and Sensors*, Springer Science & Business Media.

Note: Due to the different flying heights on aircrafts, the spatial resolution of airborne hyperspectral sensors is not fixed.

processes for hyperspectral data are Principal Component Analysis (PCA), the minimum noise fraction (MNF), a linear transformation that identifies nonorthogonal directions to maximize the signal-to-noise ratio (SNR), signal identification by minimum error (HySime) using multiple regression, and the global geometric framework for nonlinear dimensionality reduction (ISOMAP), transforming the data into a new coordinate system suiting the intrinsic shape of the data (Selva et al. 2015).

Since each substance inherits its own reflectance spectrum the sensor is able to produce a unique signature. Hyperspectral sensors are calibrated and therefore provide at-sensor radiance values. However, the pixels contain a mixture of individual spectra, which requires unmixing and subpixel analysis techniques (Gholizadeh et al. 2015). Hyperspectral image analysis relies on spectrum matching techniques, using a spectral library or field spectra and subpixel methods, which are supposed to unmix pixel information into end members. Recent methods combine pixel information with its spatial context in object-based image analysis (OBIA) (van der Meer 2006).

In terms of RSIF researchers have introduced specific pansharpening methods. They include spectral unmixing, maximum *a posteriori* (MAP) estimation, linear spectral mixture models, and Bayesian fusion. Since traditional sharpening methods are not suitable for hyperspectral data sets. Selva et al. (2015) suggest introducing a new term, namely hyper-sharpening. It refers to the fact that not only one high-resolution panchromatic band is introduced but an entire set of bands. This shows that technological developments cause requirements for new algorithms, which make RSIF a continuously evolving research field.

1.3.4 Thermal Remote Sensing

Thermal remote sensing uses sensors operating in the TIR region of the electromagnetic spectrum (cf. Figure 1.2). TIR covers the region between 3 and 35 μm in the electromagnetic spectrum. The measured element is the emitted radiation of Earth's surface as radiant temperature T_R. The principle of thermal remote sensing is depicted in Figure 1.14. Natural surfaces as well as man-made objects emit thermal radiation if their temperature is above the

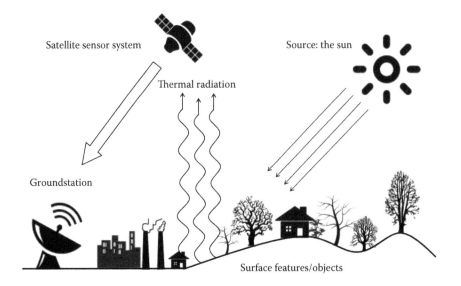

FIGURE 1.14
Principle of thermal remote sensing.

TABLE 1.6

Emissivity of Different Surfaces Using LWIR (8–14 µm)

Surface	Emissivity ε (Range 0–1)
Water	0.98
Ice	0.97–0.98
Plant leaves (healthy)	0.96–0.99
Plant leaves (dry)	0.88–0.94
Asphalt	0.96
Sand	0.93
Basalt	0.92
Granite	0.83–0.87

Source: Adapted from Lillesand, T. et al. 2015. *Remote Sensing and Image Interpretation.* Hoboken, NJ: John Wiley & Sons.

absolute zero (0 K or –273°C). The data obtained from thermal remote sensing are complementary to optical and microwave remote sensing data and therefore an important component in RSIF.

The amount of emitted radiation depends on *emissivity* ε and *kinetic temperature* T_k. Emissivity is measured as the amount of radiation that a material emits in comparison to a so-called black body. A black body is a concept that represents the ideal object that absorbs and emits all incident energy at all wavelengths (ε = 1). Examples of emissivity of different surfaces in the range of typical remote sensing TIR wavelengths of 8–14 µm are provided in Table 1.6 (Lillesand et al. 2015).

There are various defined relationships defining the emitted radiation of the black body depending on wavelength (Planck's radiation law) and temperature (Stefan–Boltzmann Law). Wien's law defines the wavelength at which maximum spectral radiance appears. The relationships are illustrated in Figure 1.15. For further reading, we refer to the book on TIR remote sensing edited by Kuenzer and Dech (2013).

The emitted radiation from Earth's surface varies with composition of the material and geometry of the observed surface. The kinetic temperature is the surface temperature of an object, measured as the amount of heat energy contained. It is influenced by two broad groups of factors, such as the heat energy budget and the thermal properties of the material (Prakash 2000) listed in Figure 1.16. It is kinetic, that is, dynamic because it is not a constant value but varies due to the factors shown.

A clear definition of TIR domains in remote sensing does not exist. There are three atmospheric windows, which allow the perception of electromagnetic radiation from the surface targets. Most popular in remote sensing is the use of 8–14 µm wavelengths TIR. 3–5 µm also belongs to TIR but are complex in interpretation since this region overlaps with solar reflection as shown in Figure 1.17. Most sensors skip the narrow ozone (O_3) absorption band (shown as dashed line in Figure 1.17).

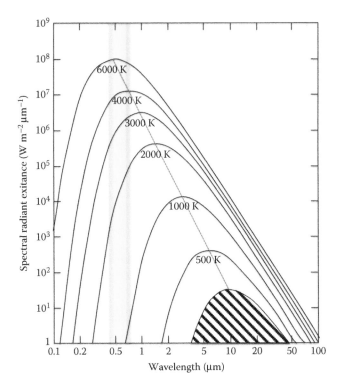

FIGURE 1.15
Relationships defining the emitted radiation of the black body expressed by the laws of Planck, Stefan–Boltzmann (area below 3000 K-curve), and Wien (red dotted line). (Adapted from Kuenzer, C. and S. Dech. 2013. *Thermal Infrared Remote Sensing*, Springer, Berlin.)

There are different modes of thermal data acquisition. They consider active versus passive, broad band versus multispectral, and daytime versus night-time acquisitions (Prakash 2000). The energy received at the sensor is transferred onto a detector using a mirror. The detectors are cooled with liquid nitrogen, which is necessary to reduce noise. Applications of thermal remote sensing include military, volcanology, glaciology, climatology, the study of surface temperature dynamics (land and sea), fire monitoring, urban heat island (UHI) mapping, and soil moisture investigations, as well as thermal water pollution. In combination with multispectral information and microwave images, object identification and the understanding of complex dynamic processes are facilitated. Due to its physically different nature, TIR data follow a different analysis than, for example reflective, optical remote sensing data. Commonly derived products are land surface temperature (LST) and sea surface temperature (SST). The detection of thermal anomalies help fire detection and monitoring or the investigation of pollution, energy leaks in buildings, and other useful information to prevent hazards and energy loss (Kuenzer and Dech 2013).

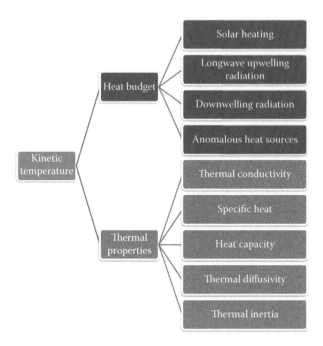

FIGURE 1.16
Components that form the kinetic temperature of an object. (Adapted from Prakash, A. 2000. *International Archives of Photogrammetry and Remote Sensing*, 33 (B1; Part 1), 239.)

There are quite a number of sensors carrying TIR acquisition capability. Common and freely accessible at a reasonable spatial resolution are Landsat TIR data. Landsat TM images contain a thermal band at 120 m spatial resolution (band 6, 10.40–12.50 µm). Landsat ETM+ provides TIR band 6 at 30 m resolution and the latest Landsat 8 Operational Land Imager (OLI) provides

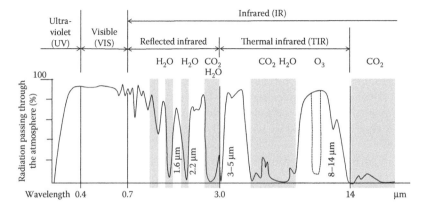

FIGURE 1.17
Atmospheric windows with reference to thermal remote sensing.

TABLE 1.7

Common Thermal Bands of Different Sensors and Their Specifications

Platform	Sensor	Spatial Resolution (m)	Band(s)	Wavelength Range (µm)
MTI	MWIR	20	J	3.50–4.10
			K	4.87–5.07
	LWIR		L	8.00–8.40
			M	8.40–8.85
			N	10.2–10.7
Landsat	TM	120	6	10.40–12.50
	ETM+	60	6	10.40–12.50
	OLI	100	TIRS 1	10.60–11.19
			TIRS 2	11.50–12.51
ASTER	TIR	90	11	8.125–8.475
			12	8.475–8.825
			13	8.925–9.275
			14	10.250–10.950
			15	10.950–11.650
MODIS[a]	TIR	1000	20	3.660–3.840
			21	3.929–3.989
			22	3.929–3.989
			23	4.020–4.080
			24	4.433–4.498
			25	4.482–4.549
AVHRR	TIR	1090	1	0.58–0.68
			2	0.725–1.00
			3A	1.58–1.64
			3B	3.55–3.93
			4	10.30–11.30
			5	11.50–12.50

[a] MODIS has 16 TIR bands in total, whereby bands 26–36 are used in cloud and ozone studies. Further details can be obtained at http://modis.gsfc.nasa.gov/about/specifications.php.

two thermal bands with 100 m spatial resolution, namely thermal infrared sensor (TIRS) 1 and 2, covering 10.60–11.19 µm and 11.50–12.51 µm, respectively. Table 1.7 provides a good overview on popular TIR sensors. It should be noted that at very low or very high temperatures, detector acquisitions saturate.

Similar to microwave remote sensing, thermal energy can be detected independent of sun illumination. Therefore, TIR provides day and nighttime data. The nighttime data become increasingly important if anomalies of heat occurrences, such as forest fires, thermal water pollution, or subsurface coal fires, are studied (Kuenzer and Dech 2013).

1.3.5 LiDAR

LiDAR is an active remote sensing method, using pulsed laser to measure distances (ranges). Similar to radar, the coherent light pulses are transmitted, reflected by a target, and detected by a receiver. The data are processed into height information providing 3-D models of Earth's surface. This is also called laser altimetry (Flood 2001) and forms the most common usage of LiDAR today. Together with the intensity of the acquired backscattered signal, surface characteristics can be imaged. Using Doppler, LiDAR velocities are obtained. LiDAR is operated on spaceborne, airborne, unmanned aerial vehicle (UAV), and terrestrial platforms. Topographic LiDAR acquires information on land surfaces based on near-infrared (NIR) laser (1.0–1.5 µm), while bathymetric LiDAR uses water-penetrating green light (0.50–0.55 µm) to map seafloor and riverbed elevations. The great advantage of LiDAR is its high accuracy and flexibility, since it can be operated any time and system parameters are adjustable, for example, flying speed/height, scan angle, pulse rate, and scan rate amongst others (Baltsavias 1999). In addition, it is capable to obtain information on different parts of the targeted surface objects based on up to six multiple pulses (Mallet and Bretar 2009). This is of great advantage in vegetation (e.g., forest: canopy, branches, trunk, ground) or urban studies to determine building heights. The data are divided into first, intermediate, and last returns and requires postprocessing. Nevertheless, it is a rather costly method and due to its operating platform, it is limited in terms of area coverage.

The acquisition of LiDAR happens through a sensor that records the backscattered light to determine the range. Together with positioning information and orientation of the platform, the so-called point clouds are generated, providing exact 3-D coordinates with intensity information. The principle of LiDAR remote sensing is illustrated in Figure 1.18.

From these clouds, reaching densities of up to 100 points per m^2, other products can be derived, such as Digital Elevation Models (DEMs), canopy, or building models, and others. Full-waveform LiDAR provides additional information about structure and physical backscattering characteristics of the targeted surface (Mallet and Bretar 2009) compared to only point cloud generations from range measurements. An illustration of the LiDAR pulse return signal is provided in Figure 1.19.

Both, SAR and LiDAR are active sensors, whereby LiDAR uses laser radiation. SAR can penetrate clouds, while thick clouds and precipitation can restrict LiDAR because of the different wavelengths. Similar to SAR, the signal is backscattered, and the energy reaching the scanner is recorded. In RSIF, it is important to understand the individual contributions of the various sensors. Table 1.8 compares the major elements of radar and LiDAR technology.

There are two types of range detection systems in operation: (1) pulsed and (2) continuous wave (CW). Pulsed LiDAR is the most popular version and transmits a series of laser pulses at 10–150 kHz pulse rate. The range is calculated from the signal travel time of the sent and received pulse and the

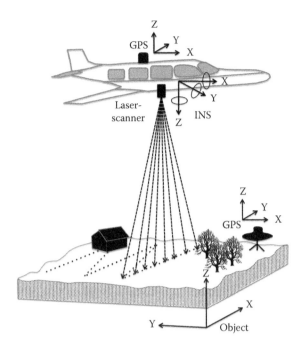

FIGURE 1.18
Principle of LiDAR remote sensing.

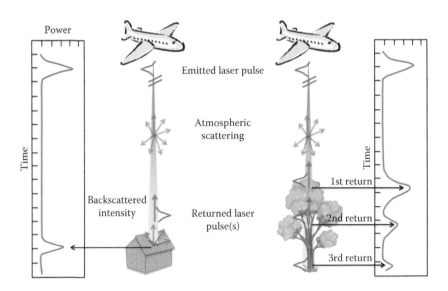

FIGURE 1.19
LiDAR pulse response signal. (Adapted from Yan, W. Y. et al. 2015. *Remote Sensing of Environment*, 158: 295–310.)

TABLE 1.8

Comparison of Radar and LiDAR Technology

Element	LiDAR	Radar
Signal	NIR, visible	Microwaves
Beam	Focused with high frequency, therefore high resolution	Width and antenna length limit spatial resolution
Look	Nadir	Side looking
Clouds	Limited by clouds and precipitation	Penetrates clouds

knowledge of the speed of light. Pulsed LiDAR systems are divided into small (0.2–3 m) and large (10–70 m) footprint types depending on platform altitude, instantaneous field of view (IFOV), and viewing angle. The different signal representation and information content for small and large footprint LiDARs are given in Figure 1.20. CW LiDAR applies a sinusoidal signal of known wavelength and determines the range from a number of full waveforms and the phase difference of the transmitted and received signal (similar to radar).

Airborne LiDAR is often mentioned as airborne laser scanning (ALS) in the literature. It reaches <0.1 m accuracy in altimetry and <0.4 cm in planimetric accuracy (Mallet and Bretar 2009). Applications include forest studies, digital elevation, and 3-D urban area modeling as well as bridge or dike monitoring, power line detection, and coastal monitoring. It also plays a major role in hazard and disaster monitoring and prevention (e.g., landslide risk analysis). Spaceborne LiDAR serves the analysis of atmospheric components and

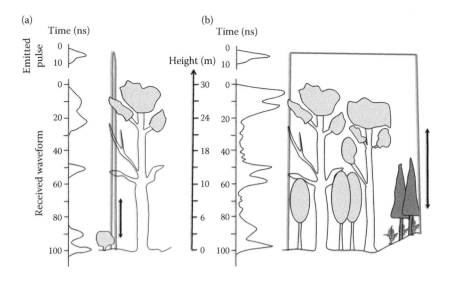

FIGURE 1.20

Multi-pulse responses with (a) small- and (b) large-footprint LiDAR. (Adapted from Mallet, C. and F. Bretar. 2009. *ISPRS Journal of Photogrammetry and Remote Sensing*, 64 (1): 1–16.)

clouds. An example is the Cloud-Aerosol LiDAR and Infrared Pathfinder Satellite Observations (CALIPSO), operating two LiDAR sensors at 532 and 1064 nm with polarization options. It has been designed to acquire high-resolution vertical profiles of aerosols and clouds.

It is rather difficult to provide a full list of LiDAR systems since there are constantly new systems emerging on the market. There are reviews in the literature containing tables listing operationally, experimentally, and commercially available systems (Baltsavias 1999; Mallet and Bretar 2009; Yan et al. 2015).

Limitations of LiDAR use are between-class spectral confusion and within-class spectral variation apart from shadowing and relief displacement (Yan et al. 2015). The combination of spectral information with ALS data significantly improves the accuracy and potential usefulness of LiDAR data, which explains the increasing interest in multi-sensor data fusion with LiDAR. Recent literature reports on successful implementation of LiDAR with other data fusion for urban areas (Chen et al. 2009; Yongmin and Yongil 2014), large area mapping (Singh et al. 2012), tree classification (Hartfield et al. 2011) just to name a few.

1.4 Image and Data Fusion

Multimodal remote sensing image and data fusion is a central research area with growing importance (Gomez-Chova et al. 2015). Earth observation satellites, airborne platforms (both manned and increasingly unmanned), and many other types of ground-based sensors, both active and passive, and covering different portions of the electromagnetic spectrum are collecting an ever increasing amount of high and higher spatial, spectral, and temporal resolution data. To analyze and fully exploit capability, vast amounts of sophisticated multi-source data requires advanced analytical and numerical image and data fusion techniques, in order to extract more meaningful, reliable, and timely results to serve the needs of the image analyst, application scientists, decision makers, and industry. In the end of the 1980s, early 1990s, remote sensing image fusion was first introduced to the remote sensing community (Welch and Ehlers 1987; Carper et al. 1990; Harris and Murray 1990; Chavez et al. 1991; Pohl 1992), and by the end of the 1990s, it had started to be accepted and widely applied by remote sensing researchers and practitioners (Pohl and Genderen 1998). Since then, it has become part of commercial software packages, product lists of satellite image providers, and an important part of processing imagery for Earth observation research. Fusion of multi-source imagery is considered as a prime solution to optimize information extraction from remote sensing data (Zhang 2010).

Over the past decade, remote sensing image fusion has developed into an operational procedure for several applications, and certain image fusion algorithms are now standardly provided in many commercial image processing software packages. However, the definition of an appropriate workflow prior to processing the imagery requires knowledge in all related fields—that is, remote sensing, image fusion, and the desired image exploitation processing and relevant application. From the results, it is can be seen that the choice of the appropriate technique, as well as the fine tuning of the individual parameters of this technique, is crucial. There is still a lack of strategic guidelines due to the complexity and variability of data selection, processing techniques, and applications. This book gives an introduction to remote sensing image fusion providing an overview on the sensors and applications. It describes data selection, application requirements, and the choice of a suitable image fusion technique. It comprises a diverse selection of successful image fusion cases that are relevant to other users and other areas of interest around the world. From these cases, common guidelines, which are valuable contributions to further applications and developments, have been derived. The availability of these guidelines will help to identify bottlenecks, further develop image fusion techniques, make best use of existing multimodal images, and provide new insights into Earth's processes. The outcome is a remote sensing image fusion textbook in which successful image fusion cases are displayed and described, embedded in common findings, and generally valid statements in the field of image fusion. The book helps newcomers to obtain a quick start into the practical value and benefits of multi-sensor image fusion. Experts will find this book useful to obtain an overview on the state of the art and understand current constraints that need to be solved in future research efforts.

Fused images may provide increased interpretation capabilities and more reliable results since data with different characteristics are combined. The images vary in spectral, spatial, and temporal resolution and therefore give a more complete view of the observed objects. For example, with high-resolution optical panchromatic image, we can see the high spatial detail of an object's shape, size, and pattern. When a multispectral image is added, we observe the spectral details of various wavelengths in various color combinations. Then, by adding a thermal infrared image of the same objects, we add temperature and emissivity of the object, making identification and understanding of the object easier. When a radar image is selected as well, further unique properties of the object of interest are provided. Additional data sources may be fused as well, such as topographic maps and thematic maps, to help understand what the object is, how it has changed over time, etc.

There are many aspects of remote sensing image fusion to consider, before one can select the optimum approach to solve a specific problem. Some key questions that need to be considered by the user are the following:

- What is the objective/application of the user? Once this is clear, then it is possible to define the necessary spectral, spatial, and temporal resolution of the remote sensing images needed to be fused.
- Which type of remote sensing imagery and data are the most useful for meeting these needs? This depends to a large degree on the

orbital characteristics of the satellite or platform characteristics of the airborne or other platform, especially in terms of imaging geometry and date and time of day of image acquisition. Weather conditions, seasonal effects, have a great influence on any fused image products. The type of landscape, especially the topography also has a major influence on the fused imagery.

- Which is the "best" technique for fusing these types of images and data to address the particular problem under consideration? This depends to a large degree on the application/problem being investigated. In addition, it is determined by the selection of appropriate interpretation methods, as when one fuses very disparate data sets such as optical data with SAR imagery, the resulting gray values do not actually refer to any physical attributes. There is still a need for ground truth for verification purposes.

- What are the necessary preprocessing steps involved? Geometry is clearly one of the major issues here, as the shape, size, and orientation of the pixels from optical satellites depend on whether the imaging platform is vertically above the object of interest, or is off-nadir, and on the precise orbital characteristics of the satellite or airborne platform. The geometry of SAR sensors is also quite different to that of optical ones, so the geometric aspects need careful consideration when fusing such images.

- Which combination of remote sensing images and other data sources are the most useful and successful? There are hundreds of studies that have been carried out to determine which combination of remote sensing images and ancillary data are best to use for a particular application. As of today, there is no one fusion method ideally suited for all applications.

In the next section, we will illustrate the need to determine objectives and application requirements for successful image fusion.

1.4.1 Objectives of Image Fusion

Each type of fusion pursues a specific objective. Image fusion is a tool to combine multi-source imagery using advanced image processing techniques. It aims at the integration of disparate and complementary data to enhance the information apparent in the images as well as to increase the reliability of the interpretation. Complementarity on the same observed area is given if data are recorded by

- Different sensors (multi-sensor)
- The same sensor scanning the same scene at different dates (multi-temporal)

- The same sensor operating in different spectral bands (multispectral)
- The same sensor at different polarizations (multi-polarization)
- The same or different sensor located on platforms flying at different heights (multi-spatial) (Simone et al. 2002)

Remote sensing image and data fusion aims to integrate the information acquired with different spatial, spectral, and temporal resolutions from sensors mounted on satellites, airborne platforms, and ground-based sensors to produce fused data that contain more detailed information than is contained in each of the individual data sources ("1 + 1 = 3") (Pohl and Genderen 1998). Hence, it increases the reliability of the interpretation, leading to more accuracy and therefore increased utility. It has also been shown that fused data provide for more robust operational performance. For example, it provides increased confidence in the interpretation result, reduced ambiguity, improved reliability, and an increase in the classification accuracy. In the end, users intend to extract useful information while avoiding artifacts reducing quality in interpretation and further processing (Laporterie and Flouzat 2003). Image fusion is applied to digital imagery for example in order to

- Sharpen images (Zhang and Mishra 2014)
- Improve geometric corrections (Toutin 2011)
- Provide stereo-viewing capabilities for stereo-photogrammetry (Toutin 2000)
- Extraction of height information (Wegner et al. 2014)
- Enhance certain features not visible in either of the single data sources alone (Robledo et al. 2009)
- Complement data sets for improved classification (Gomez-Chova et al. 2015)
- Detect changes using multi-temporal data (Bovolo and Bruzzone 2015)
- Substitute missing information (e.g., clouds-VIR, shadows-SAR) in one image with signals from another sensor image (Eckardt et al. 2013)
- Replace defective data (Shen et al. 2015)
- Provide large area coverage (Konecny 2013)

Factors to be taken into account when fusing remote sensing imagery include data accessibility, geometry, scale, spatial, spectral, temporal resolutions, and most importantly application requirements. Data accessibility depends on factors, such as geographic location, cloud cover, and available budget. Viewing geometry influences data acquisition and representation of information in the image. It is a major criterion for certain applications and influences interpretation (e.g., terrain-induced distortions) and stereo-viewing capabilities. The provision of interferometric SAR (InSAR) also requires a certain acquisition geometry. The application determines the scale at which the data are needed and directly relates to spatial, spectral, and temporal resolution.

1.4.2 Types of Fusion

One of the simplest, easiest, and the most common uses of image fusion is to increase the spatial resolution of an image. The most frequent use of this image sharpening approach is when a high-resolution panchromatic satellite or airborne image is fused with a lower resolution multispectral image (pansharpening). By means of this process, the spectral resolution of the multispectral imagery may be preserved, whilst the higher spatial resolution of the panchromatic image is incorporated, which represents the image information content of the image in much more detail. Early platforms such as SPOT2-5 (3 MS + PAN) and Landsat ETM+ (6 MS + PAN) provided a scale ratio of 2, newer systems, such as IKONOS-2, QuickBird-2, GeoEye-1/2, and Pléiades (4 MS + PAN) along with WorldView-2 (8 MS + PAN) offer a scale ratio of 4 (Selva et al. 2015) (cf. Table 1.2). Another example of image sharpening is when high-resolution optical imagery is fused with lower resolution SAR data or the other way around. This results in not only the integration of disparate data sets, but is also used to spatially enhance the imagery involved. In the mapping and especially map-updating field, the geometric accuracy and increase of spatial scales are aspects where image and data fusion techniques are applied.

In conventional photogrammetric mapping approaches, ground control points are identified in the satellite/airborne image and registered with the same feature on a topographic map. However, using image fusion, with more accurate image-to-image registration shows improved results in comparison with image-to-map registration. This is so because with image-to-map registration, there are differences in the appearance of the features used as control or tie points. The positioning of such points is greatly facilitated if both are localized in similar "views" as provided by remote sensing imagery.

An increasing number of Earth observation satellites nowadays provide 3-D stereo data sets (e.g., SPOT, ZY-3, or TH-1). Such multi-sensor data sets have been shown to be very useful in overcoming the lack of information in imagery, due, for example, to cloud cover. Hence, combinations of VIR/VIR with different spatial resolution, VIR/SAR with different incidence angles, and multiple/SAR images have been used to overcome this problem. However, several constraints should be taken into consideration, depending on the radiometric differences of the images fused to form the stereo pair.

Because of the differences in the physical nature of optical, thermal, and microwave sensor systems, fusing imagery from such different data results in enhancement of various observed features. Thus certain objects may not be easily detectable in optical imagery because spectrally similar, but they may have different temperature characteristics, and hence easily detected on fused optical/thermal images. This can be done similarly with VIR/SAR fusion. Consequently, such fused imagery makes detection, identification, and classification of certain objects much easier.

When multiple sources of image data are used in the case of classifying image, the classification accuracy is significantly improved. By fusing images from optical sensors with those taken by sensors recording in a different portion of the electromagnetic spectrum provides complimentary information that helps in discriminating certain objects. For example, in multispectral optical data, some vegetation species may not be easily classified because they may have similar spectral signatures. However, when we add a thermal image containing temperature and emissivity elements or add a microwave SAR image providing information on surface roughness, shape, and moisture content, it becomes increasingly easier to separate the different classes. Classifying fused images requires new algorithms, as the traditional statistical techniques applied to classifying solely optical imagery do not suffice for fused image data sets.

To detect changes over time, the fusion of temporal data sets from the same sensor (e.g., more than 40 years of data from Landsat) or from multiple sensors is a very useful method for such change detection of, for example, land cover. This is referred to as "temporal image fusion." Using Landsat as an example, the same place on Earth is imaged at approximately the same time of day every 16 days. The combination of these temporal images enhances the information on changes that may have occurred in the area. However, there will be seasonal differences to be taken into account, which will influence the change detection ability, unless the images are taken in the same season. With multi-sensor temporal fusion, it is almost impossible to obtain images of the same area taken at exactly the same time. Hence much emphasis has to be placed on the radiometric aspects, due to differences in atmospheric conditions, sun angle, etc. There are many different ways to do temporal image fusion. Multi-temporal SAR data provide excellent potential for change detection analysis, because of its all-weather capability. Also, the geometry of the different sensors should be carefully considered for accurate coregistration. Such radiometric and geometric preprocessing techniques are described in detail in Chapter 3.

The problem of cloud cover in optical satellite imagery is a well-known phenomenon. Especially, in the entire Tropical belt plus in temperate climatic zones, the frequent or even persistent cloud cover makes optical data almost useless or rather limited in use. In addition, the shadows of the clouds on the ground surface add to the difficulty of interpretation and classification. By fusing such cloud-covered images with SAR data, the cloudy areas can be replaced with SAR data. Alternatively, the clouds can be masked out and replaced by multi-temporal optical imagery or SAR. This RSIF solution is described and illustrated in Chapter 6. However, when replacing, clouds with SAR imagery introduces other aspects, which can cause errors. Effects such as terrain-induced geometric distortions resulting from the side-looking geometry of SAR satellites give image "gaps" or missing information in the layover and foreshortening areas as well as in the shadow areas. These gaps in SAR data can be replaced by optical imagery after accurate fusion.

Occasionally, there will be defects in the data of an image, such as loss of some scan lines, and other sensor/system errors. Again, such image defects can be removed by fusing with other images of the same area.

Apart from the levels at which fusion can be performed image fusion can be applied to various types of data sets:

1. Single sensor temporal (Amitrano et al. 2015)
2. Multi-sensor temporal (Gungor and Akar 2010)
3. Single sensor spatial (Ehlers and Klonus 2014)
4. Multi-sensor spatial (Zhang et al. 2014)
5. Single sensor product (Lu et al. 2010)
6. Single sensor angular (Longbotham et al. 2012)
7. Single sensor polarized (Du et al. 2015)
8. Remote sensing data with ancillary data (Gamba 2014)

(1) Multi-temporal image acquisition refers to the collection of data between two different overpasses of the same sensor. The temporal aspect is defined by the repeat cycle of the satellite. In early stages, two satellite sensors with the same specifications were launched at different times to ensure a continuation of a successful program (e.g., ERS, SPOT, or Landsat). Identical sensors could also be affiliated to this type of fusion because they deliver similar images but at different times. Nowadays, satellites are launched in constellations (e.g., COSMO-SkyMed or Sentinel) to enable a shorter revisit time and provide higher temporal resolution. (2) Multi-sensor multi-temporal is the largest group of fusion cases because different sensors always collect data at different dates. (3) Many optical remote sensing satellites carry two types of sensors on board, which allow the simultaneous acquisition of a high spatial resolution panchromatic band along with high spectral resolution multispectral bands. The latter are of lower spatial resolution compared to the panchromatic channel. Typical names for this type of fusion are pansharpening, spatial sharpening, or downscaling. (4) Similar to (3) the increase in spatial resolution can be performed with data from different platforms. The sharpening is performed using either a panchromatic channel from another sensor or a higher resolution SAR sensor. (5) This class is rather interesting and gained importance since the availability of multi-capability SAR. It helps enhancing certain features that are relevant to a specific application. SAR can deliver amplitude and phase as well as InSAR data. From InSAR, a DEM can be derived. Combining the intensity image with the DEM provides 3-D viewing capability. Coherence maps and InSAR deformation maps can

help natural hazard and disaster management. (6) Sensors that are able to collect high off-nadir images can "see" the same area from different orbits. Especially with the high spatial resolution of recent satellite sensors, such as WorldView-2, this opens up new perspectives by collecting multi-angular satellite images. This is of particular interest when looking at very high buildings in dense urban areas because it allows the mapping of building surfaces and roofs from all sides. Multi-angular observations also help to remove spectral distortions when used together. (7) Various spaceborne SAR sensors collect data at different polarizations. The polarized wave interacts differently with the observed objects depending on the polarization. So the different backscatter can be used for classification. (8) Combining remote sensing images with other information is common practice. This form of fusion is considered data fusion rather than image fusion and can lead to very good results. Ancillary data could be for example in situ measurements, maps, DEMs, or previous classification results. The nonimage data provide important information for the interpretation and processing of images.

The possible combinations of remote sensing data and fusion techniques are limitless as illustrated in Figure 1.21. Considering the variations in platform, sensor, processing level, fusion technique, and application parameter there is a wide range of decisions to be taken for an optimized output (Pohl and Zeng 2015).

1.4.3 Benefits

Image fusion leads to more accurate data and increased utility (Pohl and Genderen 2015). Fused data provide for robust operational performance,

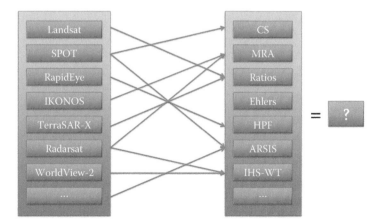

FIGURE 1.21
Illustration of variability in the types of fusion that can be performed. (Adapted from Pohl, C. and Y. Zeng. 2015. Development of a fusion approach selection tool. Paper read at *International Workshop on Image and Data Fusion*, July 21–23, 2015, at Kona, Hawaii.)

that is, increased confidence, reduced ambiguity, improved reliability, and improved classification. Image fusion aims at increasing reliability of estimation by additional information or increase interpretability by complementary information. Ideally, RSIF produces images containing reduced ambiguity, uncertainty, and errors. It should maximize the different characteristics of the input data and increase spatial, spectral, and temporal resolution while improving visual quality. It helps to increase feature recognition and classification accuracy. In a multi-sensor context, image fusion increases feasibility and reliability of information extraction and forms a very efficient approach to optimize interpretation (Li and Wang 2015). The goal of multi-sensor image fusion is the combination of complementary information in one image that serves a better understanding of the objects observed (Langford 2015). The focus of RSIF research as published in peer-reviewed international journals lies mainly on algorithm improvements and pansharpening as displayed in Figure 1.22. Increasingly, application-oriented research in RSIF is emerging.

There are two ways to consider the application of fused imagery and fused data. One is the classical Earth observation approach, looking at applications in traditional fields such as agriculture, forestry, environment, geology,

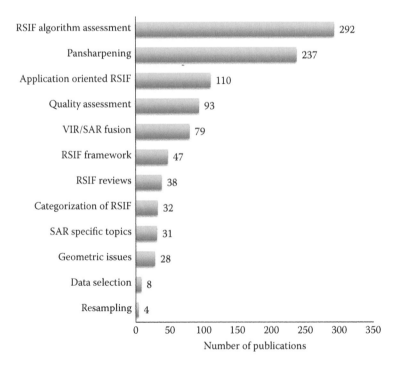

FIGURE 1.22
Focus of published RSIF research.

mineral exploration, land use, land cover, urban, disasters, and many others. The application of remote sensing image fusion is the topic of Chapter 6, where many practical examples and case studies are presented. Another approach is to look at the applications from a technology point of view. Hence, fusion techniques have shown to be useful in applications, such as object detection, recognition, identification, and classification, to object tracking, change detection, and decision making. This is why it has been successfully applied to Earth observation domain, computer vision, medical image analysis, and of course to many military and security topics.

1.4.4 Limitations

With all the advantages of accessing multiple sensor data in an integrated fashion, there are limitations to this approach. One of these limitations refers directly to the capability of the individual sensor. High-resolution optical sensors provide more details. However, they are often limited in bandwidth and in the number of bands. The signal acquired by SAR provides complementary information to VIR data but mixes geometric and physical object properties. Hyperspectral data provide a wide spectral profile but lacks spatial detail and leads to mixed pixel information. Sensors that acquire 3-D data, such as stereo mapping satellites, InSAR, LiDAR, and stereo photogrammetry deliver a spatial resolution that is complex to align with other sensors (Salentinig and Gamba 2015).

1.5 Definitions and Terminology for This Book

Even though most of the terminology is common in remote sensing, different authors describe the same object with different words. In order to facilitate a proper understanding and to avoid confusion, we explain a few reoccurring terms utilized in this book. A general definition of image fusion is given as "Image fusion is the combination of two or more different images to form a new image by using a certain algorithm" (Genderen and Pohl 1994). The broader topic of *data fusion* is a process dealing with data and information from multiple sources to achieve refined/improved information for decision making (Hall and McMullen 2004). In the context of remote sensing, Mangolini (1994) defined it as a group of methods and approaches using multisource data of different nature to increase the quality of information contained in the data. The input images or data sources are different; not only spatially, spectrally, and temporally, but may also include data sources such as topographic maps, GPS coordinates, geophysical information, voice, and text data.

In terms of information fusion, the most common application in the remote sensing field is where the fused images have already been interpreted and analyzed to reach the information required, and that this information is then fused with other types of nonremote sensing data to provide relevant reliable information to the decision maker. This book does not treat information fusion, but focuses on remote sensing image fusion. For those readers interested in information fusion, the following publications are recommended for further reading (Waltz and Llinas 1990; Zhang 2010; Khaleghi et al. 2013).

Many other terms have been used in the literature. These include the following:

- Data integration (Bahiru and Woldai 2016)
- Data merging (Minghelli-Roman et al. 2006)
- Data matching (Nikolakopoulos 2008)
- Data synergy (Vanhellemont et al. 2014)
- Data combination (Torabzadeh et al. 2014)

However, since the publication of the first review article (Pohl and Genderen 1998) and the work of the Data Fusion Interest Group (Wald 2002, 2001), the term fusion is the most common to be found in the literature.

1.5.1 Remote Sensing Images

A *platform* (satellite, airplane, UAV, and terrestrial platform) can carry several sensors with different specifications. One *sensor* can produce multiple *bands* or *channels* that form one *image*. Therefore, remote sensing images contain one or more *bands* that are also called *channels*. In the book, we use both terms equally.

In the optical case (visible and infrared spectrum), including hyperspectral sensors, different observed objects can be identified by their *spectral signature*, which varies due to the different absorption and reflection. It is made up of the *radiance* value in the different bands. Radiance is the radiant flux emitted, reflected, transmitted, or received by a surface. *Spectral radiance* is the radiance of a surface per unit frequency or wavelength, which makes up the signature.

A special case is thermal remote sensing. A thermal sensor collects the relative differences in emitted thermal energy from Earth's surface using the TIR region of the electromagnetic spectrum. The information provided is the *thermal radiation*, which occurs through solar radiation and partly through internal Earth heat flux. It depends on the *emissivity* of the observed surface. Measures of relevance in the context of thermal remote sensing are SST and LST.

In the case of SAR, the sensor usually acquires one type of information related to the *backscatter* behavior of the observed objects. The amount of energy backscattered from the ground depends on the signal (wavelength and incidence angle) and the observed object (dielectric properties, surface roughness, and terrain elevation). The original measurement, however, includes, apart from the signal travel time, phase and intensity information.

LiDAR scans Earth's surface using laser and provides 3-D coordinates of points collected in a so-called *point cloud*. In addition, surface characteristics can be studied based on new-infrared laser operating at 1064 or 1550 nm. Last but not least, LiDAR can deliver *peak laser energy* (intensity) images similar to SAR so that backscattered signals can be analyzed to understand ground cover.

1.5.2 Resolution

Resolution is a major player in RSIF. Basically, we are dealing with four types of resolution, as shown in Figure 1.23:

1. Spatial resolution
2. Spectral resolution
3. Radiometric resolution
4. Temporal resolution

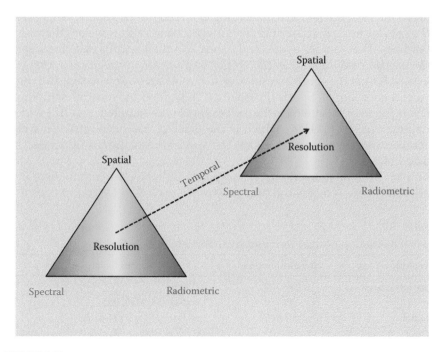

FIGURE 1.23
Types of resolution in remote sensing.

The *spatial resolution* refers to the ground sampling distance (GSD) of a sensor. It should not be confused with *pixel resolution*, which is often an artificial value to create regular numbers and depends on the resampling. An example is RapidEye-5, which provides 6.5-m GSD but is provided in 5-m pixel resolution. It is sometimes necessary in image fusion to up or down sample an image to match with another image. This is the artificial pixel resolution and does not affect the spatial resolution of the sensor itself. The literature provides different definitions whether a data set is of high or low resolutions. Terms like *very high* or *coarse* resolution are popular. We have tried to homogenize the terminology by providing an overview in Table 1.9. It should be considered that this is not a generic definition but overly valid for current satellite systems. It is obvious that the characterization has shifted with the increase in spatial resolution over the years. What used to be *high* resolution, for example, SPOT-1, 2, and 3 at 10 m became *medium* resolution now because we have sensors, such as WorldView-2, providing submeter resolution.

A matter that is closely related to spatial resolution is the *scale* at which the information extracted from remote sensing data can be represented. As does the spatial resolution, it refers to the GSD that a sensor is capable of providing. In the context of RSIF, the scale ratio is an important aspect in the data and algorithm selection as well as in the application itself. A small ratio requires sophisticated processing to account for the large discrepancy in detail that is contained in a pixel (Ling et al. 2008; Ehlers and Klonus 2014).

The *spectral resolution* describes the different bands and wavelengths a sensor utilizes. The number of spectral bands and their widths describe a sensor's spectral resolution. It is the ability of a sensor to specify wavelength intervals. A finer the spectral resolution means a narrower wavelength-range for a certain channel. Spectral resolution is defined by spectral sampling, which is the interval at which the collected data are sampled, and full-width at half-maximum (FWHM), which is the detector response retrieved from calibration (Borengasser et al. 2008). The spectral resolution is illustrated in Figure 1.24.

Similar to the relationship between spatial and pixel resolution, there is a relationship between spectral and *radiometric resolution*. Digital images are

TABLE 1.9

Spatial Resolution-Related Terminology in Satellite Remote Sensing

Resolution Term	Resolution (m)	Sensor Examples
Very low/coarse	250–1000	MODIS
Low	30–250	Landsat TM, MSS, ERS
Medium	10–30	Landsat ETM+, SPOT-4, IRS-3
High	2–10	IKONOS, SPOT-5, RapidEye, Formosat-2
Very high	<1–2	QuickBird, WorldView-2, GeoEye, Radarsat-2, TerraSAR-X

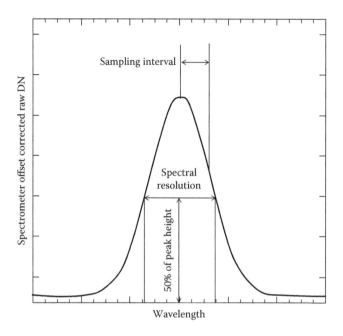

FIGURE 1.24
Relationships of wavelength, sampling distance, and FWHM for the description of the spectral resolution of a sensor. (Adapted from Borengasser, M. et al. 2008. Hyperspectral remote sensing: Principles and applications. In *Remote Sensing Applications*, edited by Q. Weng. Boca Raton, FL: CRC Press.)

stored using a certain number of bits representing the energy recorded by the sensor. Two to the power of bit equal the number of gray scale values. As an example, 8 bits recording results in 256 digital values available for coding. This means the image values are stored in gray values ranging from 0 to 255. The higher the bits, the more gray scale values can be differentiated by a certain sensor. Figure 1.25 provides the calculation and illustration of radiometric resolution.

Temporal resolution refers to the revisit cycle of the platform, meaning how much time elapses between two acquisitions over the same area on Earth.

Bits	Range of values	Gray values	
1 bit	$2^1 = 2$ (0–1)	0	1
4 bits	$2^4 = 16$ (0–15)	0	15
8 bits	$2^8 = 256$ (0–255)	0	255

FIGURE 1.25
Calculation and illustration of the radiometric resolution.

In the description of satellite imagery, it is called repeat cycle. Modern satellite systems provide constellations of satellites carrying a similar sensor on board to increase revisit time and therefore the temporal resolution of a sensor. Examples are COSMO-SkyMed, Sentinel, and the future Radarsat Constellation Mission (RCM). Another option to reduce the revisit time is a sensor's capability to acquire off nadir, like for example SPOT or many high-resolution satellites, although such off-nadir viewing capability requires the image to be resampled. Varying pixel size at nadir and at the off-angle produces the need for providing equal pixel spacing prior to image fusion.

1.6 Summary

In this introductory chapter, we have first of all explained the structure of this book. Then, before describing the purpose and objectives of remote sensing image fusion, we have given a brief introduction to the principles, techniques, and applications of remote sensing. This includes a discussion on optical, thermal infrared, radar, hyperspectral, and LiDAR systems. The chapter then explained the main objectives of RSIF, and illustrated the types of remote sensing imagery commonly fused. The many factors to be taken into account prior to fusing the imagery are described, and the main limitations and the benefits of RSIF are explained. The next chapter introduces the various levels of image and data fusion to put RSIF in the right context.

References

Amitrano, D., G. Di Martino, A. Iodice, D. Riccio, and G. Ruello. 2015. A new framework for SAR multitemporal data RGB representation: Rationale and products. *Geoscience and Remote Sensing, IEEE Transactions on* 53 (1):117–133.

Bahiru, E. A. and T. Woldai. 2016. Integrated geological mapping approach and gold mineralization in Buhweju area, Uganda. *Ore Geology Reviews* 72 (Part 1): 777–793.

Bakker, W. H., Feringa, W., Gieske, A. S. M. et al. 2009. *Principles of Remote Sensing, ITC Educational Textbook Series 2*. edited by K. Tempfli, N. Kerle, G. C. Huurneman, and L. L. F. Janssen. Enschede: ITC, International Institute for Aerospace Survey and Earth Sciences, 591 pp.

Baltsavias, E. P. 1999. Airborne laser scanning: Existing systems and firms and other resources. *ISPRS Journal of Photogrammetry and Remote Sensing* 54 (2–3):164–198.

Bannon, D. 2009. Hyperspectral imaging: Cubes and slices. *Nature Photonics* 3 (11):627–629.

Borengasser, M., W. S. Hungate, and R. Watkins. 2008. *Hyperspectral Remote Sensing: Principles and Applications*. Boca Raton, FL: CRC Press, 128 pp.

Bovolo, F. and L. Bruzzone. 2015. The time variable in data fusion: A change detection perspective. *Geoscience and Remote Sensing Magazine, IEEE* 3 (3):8–26.

Campbell, J. B. and R. H. Wynne. 2011. *Introduction to Remote Sensing*. New York: Guilford Press.

Carper, W. J., T. M. Lillesand, and P. W. Kiefer. 1990. The use of intensity-hue-saturation transformations for merging spot panchromatic and multispectral image data. *Photogrammetric Engineering & Remote Sensing* 56 (4):459–467.

Chavez, P. S., S. C. Sides, and J. A. Anderson. 1991. Comparison of three different methods to merge multiresolution and multispectral data: Landsat TM and SPOT panchromatic. *Photogrammetric Engineering & Remote Sensing* 57 (3):295–303.

Chen, Y., W. Su, J. Li, and Z. Sun. 2009. Hierarchical object oriented classification using very high resolution imagery and LIDAR data over urban areas. *Advances in Space Research* 43 (7):1101–1110.

Chuvieco, E. and A. Huete. 2009. *Fundamentals of Satellite Remote Sensing*. Boca Raton, FL: CRC Press.

Ciampalini, A., F. Andre, F. Garfagnoli et al. 2015. Improved estimation of soil clay content by the fusion of remote hyperspectral and proximal geophysical sensing. *Journal of Applied Geophysics* 116:135–145.

Dallemand, J., J. Lichteneger, R. Raney, and R. Schumann. 1993. Radar imagery: Theory and application. In *FAO Lecture Notes*. Rome, Italy.

Dopheide, E., F. van der Meer, R. Sliuzas et al. 2013. *The Core of GIScience: A Systems-Based Approach*. Enschede: ITC.

Dor, E. B., T. Malthus, A. Plaza, and D. Schläpfer. 2013. Hyperspectral remote sensing. In *Airborne Measurements for Environmental Research: Methods and Instruments*, edited by M. Wendisch and J.-L. Brenguier. Hoboken, NJ: John Wiley & Sons, pp. 419–465.

Du, P., A. Samat, B. Waske, S. Liu, and Z. Li. 2015. Random forest and rotation forest for fully polarized SAR image classification using polarimetric and spatial features. *ISPRS Journal of Photogrammetry and Remote Sensing* 105:38–53.

Eckardt, R., C. Berger, C. Thiel, and C. Schmullius. 2013. Removal of optically thick clouds from multi-spectral satellite images using multi-frequency SAR data. *Remote Sensing* 5 (6):2973.

Ehlers, M. and S. Klonus. 2014. Scale issues in multisensor image fusion. In *Scale Issues in Remote Sensing*, edited by Qihao Weng. Noboken, NJ: John Wiley & Sons, Inc, pp. 13–33.

Flood, M. 2001. Laser altimetry: From science to commercial LIDAR mapping. *Photogrammetric Engineering and Remote Sensing* 67 (11):1209–1217.

Gaber, A., F. Soliman, M. Koch, and F. El-Baz. 2015. Using full-polarimetric SAR data to characterize the surface sediments in desert areas: A case study in El-Gallaba Plain, Egypt. *Remote Sensing of Environment* 162:11–28.

Gamba, P. 2014. Image and data fusion in remote sensing of urban areas: Status issues and research trends. *International Journal of Image and Data Fusion* 5 (1):2–12.

Genderen, J. L. van and C. Pohl. 1994. Image fusion: Issues, techniques and applications. *Intelligent Image Fusion, Proceedings EARSeL Workshop*, Strasbourg, France, 11:18–26.

Gholizadeh, H., B. Mojaradi, and M. J. V. Zoej. 2015. Local prototype space-based band selection for hyperspectral subpixel analysis. *Photogrammetrie-Fernerkundung-Geoinformation* 2015 (5):373–380.

Goetz, A. F. H., G. Vane, J. E. Solomon, and B. N. Rock. 1985. Imaging spectrometry for earth remote sensing. *Science* 228 (4704):1147–1153.

Gomez-Chova, L., D. Tuia, G. Moser, and G. Camps-Valls. 2015. Multimodal classification of remote sensing images: A review and future directions. *Proceedings of the IEEE* 103 (9):1560–1584.

Gungor, O. and O. Akar. 2010. Multi sensor data fusion for change detection. *Scientific Research and Essays* 5 (18):2823–2831.

Hall, D. L. and S. A. H. McMullen. 2004. *Mathematical Techniques in Multisensor Data Fusion*. Norwood, MA: Artech House.

Harris, J. R. and R. Murray. 1990. IHS transform for the integration of radar imagery with other remotely sensed data. *Photogrammetric Engineering and Remote Sensing* 56 (12):1631–1641.

Hartfield, K. A., K. I. Landau, and W. J. D. v. Leeuwen. 2011. Fusion of high resolution aerial multispectral and LiDAR data: Land cover in the context of urban mosquito habitat. *Remote Sensing* 3 (11):2364.

Kaewpijit, S., J. Le Moigne, and T. El-Ghazawi. 2003. Automatic reduction of hyperspectral imagery using wavelet spectral analysis. *Geoscience and Remote Sensing, IEEE Transactions on* 41 (4):863–871.

Khaleghi, B., A. Khamis, F. O. Karray, and S. N. Razavi. 2013. Multisensor data fusion: A review of the state-of-the-art. *Information Fusion* 14 (1):28–44.

Konecny, G. 2013. The International Society for Photogrammetry and Remote Sensing (ISPRS) study on the status of mapping in the world. Paper read at International Workshop on Global Geospatial Information.

Kramer, H. J. 2012. *Observation of the Earth and Its Environment: Survey of Missions and Sensors*. Berlin: Springer Science & Business Media.

Kuenzer, C. and S. Dech. 2013. *Thermal Infrared Remote Sensing*. Berlin: Springer.

Langford, R. L. 2015. Temporal merging of remote sensing data to enhance spectral regolith, lithological and alteration patterns for regional mineral exploration. *Ore Geology Reviews* 68:14–29.

Laporterie, F. and G. Flouzat. 2003. The morphological pyramid concept as a tool for multi-resolution data fusion in remote sensing. *Integr. Comput.-Aided Eng.* 10 (1):63–79.

Li, X. and L. Wang. 2015. On the study of fusion techniques for bad geological remote sensing image. *Journal of Ambient Intelligence and Humanized Computing* 6 (1):141–149.

Lillesand, T., R. W. Kiefer, and J. Chipman. 2015. *Remote Sensing and Image Interpretation*. Hoboken, NJ: John Wiley & Sons.

Ling, Y., M. Ehlers, E. L. Usery, and M. Madden. 2008. Effects of spatial resolution ratio in image fusion. *International Journal of Remote Sensing* 29 (7):2157–2167.

Longbotham, N., C. Chaapel, L. Bleiler, C. Padwick, W. J. Emery, and F. Pacifici. 2012. Very high resolution multiangle urban classification analysis. *Geoscience and Remote Sensing, IEEE Transactions on* 50 (4):1155–1170.

Lu, Z., D. Dzurisin, H.-S. Jung, J. Zhang, and Y. Zhang. 2010. Radar image and data fusion for natural hazards characterisation. *International Journal of Image and Data Fusion* 1 (3):217–242.

Mallet, C. and F. Bretar. 2009. Full-waveform topographic lidar: State-of-the-art. *ISPRS Journal of Photogrammetry and Remote Sensing* 64 (1):1–16.

Mangolini, M. 1994. Benefit of fusion of multisensor images in remote sensing and photo-interpretation. Dissertation, Centre Énergétique et Procédés (CEP), Université de Nice, Sophia-Antipolis, France.

Manolakis, D., C. Siracusa, and G. Shaw. 2001. Hyperspectral subpixel target detection using the linear mixing model. *Geoscience and Remote Sensing, IEEE Transactions on* 39 (7):1392–1409.

Minghelli-Roman, A., L. Polidori, S. Mathieu-Blanc, L. Loubersac, and F. Cauneau. 2006. Spatial resolution improvement by merging MERIS-ETM images for coastal water monitoring. *IEEE Geoscience and Remote Sensing Letters* 3 (2): 227–231.

Nikolakopoulos, K. G. 2008. Comparison of nine fusion techniques for very high resolution data. *Photogrammetric Engineering & Remote Sensing* 74 (5):647–659.

Pohl, C. 1992. Data Integration Techniques. *EARSeL Workshop on Analysis of Earth Observation Space Data Integration*, October 28–30, 2015, at Sinaia, Romania, pp. 8–13.

Pohl, C. and J. L. van Genderen. 1998. Review article multisensor image fusion in remote sensing: Concepts, methods and applications. *International Journal of Remote Sensing* 19 (5):823–854.

Pohl, C. and J. L. van Genderen. 2015. Structuring contemporary remote sensing image fusion. *International Journal of Image and Data Fusion* 6 (1):3–21.

Pohl, C. and Y. Zeng. 2015. Development of a fusion approach selection tool. Paper read at *International Workshop on Image and Data Fusion*, July 21–23, 2015, at Kona, Hawaii.

Prakash, A. 2000. Thermal remote sensing: Concepts, issues and applications. *International Archives of Photogrammetry and Remote Sensing* 33 (B1; Part 1):239–243.

Robledo, L., M. Carrasco, and D. Mery. 2009. A survey of land mine detection technology. *International Journal of Remote Sensing* 30 (9):2399–2410.

Salentinig, A. and P. Gamba. 2015. Combining SAR-based and multispectral-based extractions to map urban areas at multiple spatial resolutions. *Geoscience and Remote Sensing Magazine, IEEE* 3 (3):100–112.

Schowengerdt, R. A. 2007. Preface to the third edition. In *Remote Sensing*, 3rd ed., edited by R. A. Schowengerdt. Burlington: Academic Press.

Schreier, G. 1993. *SAR Geocoding: Data and Systems*. Berlin: Wichmann.

Selva, M., B. Aiazzi, F. Butera, L. Chiarantini, and S. Baronti. 2015. Hyper-sharpening: A first approach on SIM-GA data. *IEEE Journal of Selected Topics in Applied Earth Observations and Remote Sensing* 8 (6):3008–3024.

Shen, H., X. Li, Q. Cheng et al. 2015. Missing information reconstruction of remote sensing data: A technical review. *Geoscience and Remote Sensing Magazine, IEEE* 3 (3):61–85.

Simone, G., A. Farina, F. C. Morabito, S. B. Serpico, and L. Bruzzone. 2002. Image fusion techniques for remote sensing applications. *Information Fusion* 3 (1):3–15.

Singh, K. K., J. B. Vogler, D. A. Shoemaker, and R. K. Meentemeyer. 2012. LiDAR-Landsat data fusion for large-area assessment of urban land cover: Balancing spatial resolution, data volume and mapping accuracy. *ISPRS Journal of Photogrammetry and Remote Sensing* 74:110–121.

Thenkabail, P. S. (eds.). 2015. *Remote Sensing Handbook—Three Volume Set*. Boca Raton, FL: CRC Press, 2200 pp.

Torabzadeh, H., F. Morsdorf, and M. E. Schaepman. 2014. Fusion of imaging spectroscopy and airborne laser scanning data for characterization of forest ecosystems—A review. *ISPRS Journal of Photogrammetry and Remote Sensing* 97:25–35.

Toutin, T. 2000. Stereo-mapping with SPOT-P and ERS-1 SAR images. *International Journal of Remote Sensing* 21 (8):1657–1674.

Toutin, T. 2011. State-of-the-art of geometric correction of remote sensing data: A data fusion perspective. *International Journal of Image and Data Fusion* 2 (1):3–35.

Touzi, R., W. M. Boerner, J. S. Lee, and E. Lueneburg. 2004. A review of polarimetry in the context of synthetic aperture radar: Concepts and information extraction. *Canadian Journal of Remote Sensing* 30 (3):380–407.

van der Meer, F. 2006. The effectiveness of spectral similarity measures for the analysis of hyperspectral imagery. *International Journal of Applied Earth Observation and Geoinformation* 8 (1):3–17.

Vanhellemont, Q., G. Neukermans, and K. Ruddick. 2014. Synergy between polar-orbiting and geostationary sensors: Remote sensing of the ocean at high spatial and high temporal resolution. *Remote Sensing of Environment* 146:49–62.

Wald, L. 2001. The present achievements of the EARSeL SIG Data Fusion. Paper read at *EARSeL Symposium 2000 "A Decade of TransEuropean Remote Sensing Cooperation,"* Dresden, Germany.

Wald, L. 2002. Data fusion. Definitions and architectures. Fusion of images of different spatial resolutions. http://www.ebookdb.org/reading/2FGD1E4422461C3 83E5F7F69/Data-Fusion—Definitions-And-Architectures—Fusion-Of-Images-Of-Different-Spatia; http://www.ensmp.fr/Presses.

Waltz, E. L. and J. Llinas. 1990. *Multisensor Data Fusion.* Norwood, MA: Artech House, Inc.

Wegner, J. D., J. R. Ziehn, and U. Soergel. 2014. Combining high-resolution optical and InSAR features for height estimation of buildings with flat roofs. *Geoscience and Remote Sensing, IEEE Transactions on* 52 (9):5840–5854.

Welch, R. and M. Ehlers, 1987. Merging SPOT HRV and Landsat TM Data. *Photogrammetric Engineering and Remote Sensing* 53 (3):301–303.

Yan, W. Y., A. Shaker, and N. El-Ashmawy. 2015. Urban land cover classification using airborne LiDAR data: A review. *Remote Sensing of Environment* 158:295–310.

Yongmin, K. and K. Yongil. 2014. Improved classification accuracy based on the output-level fusion of high-resolution satellite images and airborne LiDAR data in urban area. *Geoscience and Remote Sensing Letters, IEEE* 11 (3):636–640.

Zhang, H., B. Huang, and L. Yu. 2014. Intermodality models in pan-sharpening: Analysis based on remote sensing physics. *International Journal of Remote Sensing* 35 (2):515–531.

Zhang, J. 2010. Multi-source remote sensing data fusion: Status and trends. *International Journal of Image and Data Fusion* 1 (1):5–24.

Zhang, Y. and R. K. Mishra. 2014. From UNB PanSharp to Fuze Go—The success behind the pan-sharpening algorithm. *International Journal of Image and Data Fusion* 5 (1):39–53.

2

Fusion Levels

Image fusion forms a subgroup of data fusion and uses pixel-based methods. It is very popular in remote sensing for two reasons: (1) Remote sensing acquires diverse, complementary images, and (2) RSIF ensures the combination of the data with the least alteration of their values. This chapter positions pixel-based methods within data fusion, in the neighborhood of feature- and decision-based methods. Originally, three levels were established in the remote sensing community, that is, the pixel, feature, and decision levels. In the meantime, a fourth level, namely, the subpixel level has been introduced. Some researchers include the original *signal* or in other words *sensor* as the fourth level as shown in Figure 2.1. It is defined as "the problem of combining multiple measurements from sensors into a single measurement of the sensed object or attribute called the parameter" (McKendall and Mintz 1988). With a quality requirement, the definition is given as: "Sensor fusion is the combining of sensory data or data derived from sensory data such that the resulting information is in some sense better than would be possible when these sources were used individually" (Elmenreich 2002). In remote sensing, signal fusion does not play a major role because the signal is converted into values and images for further consideration. All other levels have their importance in remote sensing. The categorization is necessary to understand remote sensing image fusion as a whole. It is meant to help the reader obtain an idea of the individual characteristics and concerns in each level. In reality, levels are often combined into hybrid fusion because it is most beneficial. In addition, it is sometimes rather difficult to assign a certain approach to one level. It has to be stated that this book was written with a focus on remote sensing *image* fusion, which refers to the *pixel level*. The remaining chapters in the book will continue to refer to the levels where appropriate. Last but not least, this chapter contains a section on the value of image and data fusion using remote sensing with other data such as information from a geographic information system, Internet resources, ground truth, and other observations, containing references to higher levels of fusion in remote sensing. The details on the different techniques used in remote sensing image fusion will be described in Chapter 4.

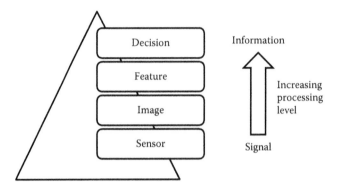

FIGURE 2.1
Definition of fusion levels and their position in the processing chain.

2.1 Data Fusion

The first definition on data fusion can be found in the Data Fusion Lexicon (White 1987), which describes data fusion as, "A process dealing with the association, correlation, and combination of data and information from single and multiple sources to achieve refined position and identity estimates, and complete and timely assessments of situations and threats as well as their significance." This definition shows that data fusion encompasses image, feature, and information fusion levels (see Figure 2.2). The challenge is to find the optimum combination, procedure, and therefore information to achieve the best decision.

Waltz and Llinas (1990) refined the definition providing the objectives of data fusion "to combine elements of raw data from different sources into a single set of meaningful information that is of greater benefit than the sum of the contributing parts." Data fusion provides the means to benefit from all four levels, which is the trend in remote sensing. The value is expressed in a definition provided by Paradis et al. (1997). They state, "Data fusion is an adaptive information process that continuously transforms available data and information into richer information, through continuous refinement of hypotheses or inferences about real-world events." A Special Interest Group on Data Fusion initiated the first comprehensive discussion in remote sensing. They define data fusion as "a formal framework, in which are expressed means and tools for the alliance of data originating from different sources. It aims at obtaining information of greater quality; the exact definition of 'greater quality' will depend upon the application" (Wald 1998). This is reflected in our definition of the levels in remote sensing data fusion and is depicted in Figure 2.2.

In the literature, the terms "information fusion" and "data fusion" are used synonymously. For this book, we clearly distinguish between data fusion, as

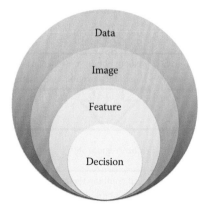

FIGURE 2.2
Relationship of the different fusion levels and their definition within; data fusion comprises all levels, while image fusion forms a subgroup comprising feature, and then decision fusion. The latter two levels are relevant after further processing the images to extract features and information, respectively.

the overall term comprising three further processed levels of data, namely, images, features, and information. With each increased processing level, we enter higher semantic levels (Castanedo 2013). Images can be fused with all kinds of other data, such as maps, geographic information system (GIS) layers, models, or statistics. In this book, we consider image with image fusion in particular. In Chapters 6 and 7, we will briefly discuss data fusion examples and issues.

2.2 Fusion in Remote Sensing

In the context of fusion levels, the terminology used in the literature is not unified. Pixel level is also referred to as *iconic* or *image* level. Other expressions for feature level are *object* or *regional* level. Decision level also appears as *information* or *symbolic* level in publications. Mostly the terminology is adapted to the application field for which the fusion process is applied. Sometimes it depends on grown structures in certain research groups. The most established terms for the levels are the following:

- Image fusion
- Feature fusion
- Decision fusion

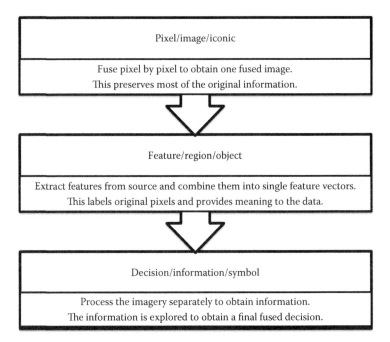

FIGURE 2.3
Terminology used for RSIF-level definitions.

Figure 2.3 provides an overview on the terminology and the definitions as they appear in the literature. The most established and cited definitions are based on the first comprehensive RSIF review (Pohl and Genderen 1998). The advancement in sensor technology has led to an increase in spatial resolution. This has introduced a fourth fusion level that is gaining more and more attention: the subpixel level (Gomez-Chova et al. 2015). This level is also considered in the following descriptions, even though the main emphasis of this book obviously lies on RSIF.

Undoubtedly, each individual level requires different processing steps. The levels are distinguished by where the actual fusion step takes place (see Figure 2.4). Image fusion requires sensor-specific corrections and coregistration prior to fusion. For feature fusion, objects of interest are extracted from the image data and fused at feature level. In decision fusion, the features are labeled to produce information, which is then fused.

All this is embedded in the entire data to the information processing chain as shown in Figure 2.5. The signal or raw data are acquired by the satellite, downlinked to the ground station, and then processed. Next the data are corrected for sensor and environment induced factors, such as sensor geometry and terrain geometry, atmosphere, etc. The result of the preprocessing step is calibrated and located image data that can be fed into the fusion processed or further refined into features or information. The final output is the fused

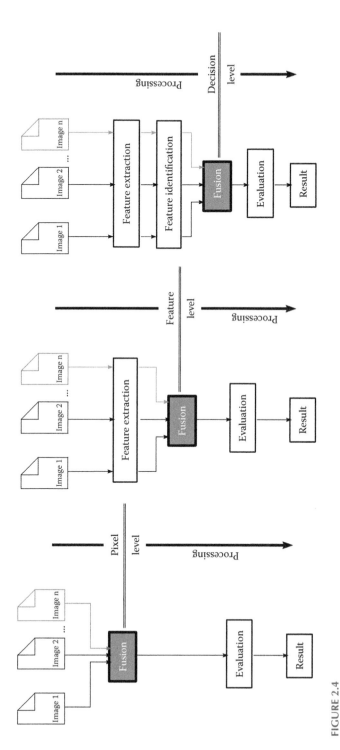

FIGURE 2.4

General processing steps for different fusion levels. (Adapted from Pohl, C. and J. L. van Genderen. 1998. *International Journal of Remote Sensing* 19(5): 823–854.)

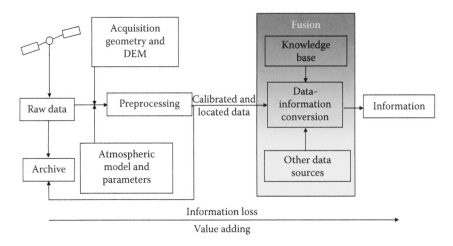

FIGURE 2.5
Larger fusion context and process from image to information conversion. (Adapted from MacDonald, R. 1997. From space data to information. Paper read at *Joint ISPRS Working Groups I/1, I/3, IV/4 Workshop on Sensors and Mapping from Space*, September 29–October 2, 1997, Hannover, Germany.)

product, which could be an image, a feature, or a decision. From data acquisition to the final output, we observe an information loss but we add value.

2.3 Subpixel Level

Subpixel is a concept that is based on the fact that the signal represented in digital numbers per pixel originates from different objects. The signal response collected in one pixel is heterogeneous, meaning that signals from various physical sources are mixed (Farah and Ahmed 2010). Mixed pixels can lead to missing out important information on land cover (Delalieux et al. 2014). In the fusion process, higher spatial resolution data are used to "unmix" lower-resolution pixels, which usually contain higher resolution in the spectral or temporal domain. Subpixel mapping (SPM) produces higher spatial resolution classification than the input coarse spatial resolution image would deliver (Qunming et al. 2015).

SPM plays a major role in change detection and will be discussed technique-wise in Chapter 4 (spatio-temporal fusion) and application-wise in Chapter 6 (change detection) in more detail. The concept of subpixel fusion is a relevant issue in change detection, in particular in the use of high temporal but low spatial resolution imagery from sensors like the moderate-resolution imaging spectrometer (MODIS) or the medium-resolution imaging spectro-radiometer (MERIS) and low temporal but higher spatial resolution data like

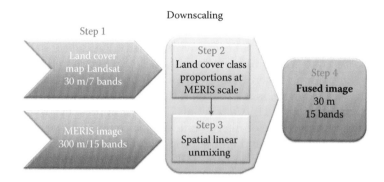

FIGURE 2.6
Spatial unmixing scheme for spatio-temporal fusion at subpixel level. (Adapted from Amoros-Lopez, J. et al. 2011. *IEEE Geoscience and Remote Sensing Letters* 8(5): 844–848.)

Landsat TM or SPOT. The algorithms differ in the type of resampling used to increase spatial resolution to subpixel size using a higher-resolution type data. Often they are iterative processes that are computer powered and time consuming but effective. Terms of relevance in the context of subpixel fusion are spectral unmixing, SPM, and super-resolution mapping (SRM) (Qunming et al. 2015). RSIF is applied using multiresolution and multi-source spatial unmixing to overcome the cloud cover problem of optical remote sensing. In the example depicted in Figure 2.6, the process derives land cover classes from the high spatial resolution image of which the proportion matrix for each MERIS pixel delivers posterior probabilities at subpixel level. The spatial unmixing is performed on the MERIS image followed by the actual fusion where the high spatial resolution fused pixels are formed from a linear combination of the estimated MERIS class weighted by their Landsat membership (Amoros-Lopez et al. 2011).

2.4 Pixel Level

Image fusion is a process whereby the combination of multiple input images produces a single composite containing the information of the input images (Mitchell 2010). Multimodal image fusion has the capability to integrate complementary information as well as provide different views of Earth's surface, which helps the understanding of complex processes. The objective of creating a fused image is to improve the quality of the information contained in the output image. The process is called synergy. Image fusion at pixel level means fusion at the lowest processing level referring to the merging of measured physical parameters. It uses raster data that is at least coregistered but most commonly geocoded. The coregistration and geocoding plays an essential role because misregistration causes artificial colors or features in multi-sensor data sets, which falsify the interpretation later on. RSIF techniques have different sensitivities to misregistration depending on the approach chosen. The improved Gram–Schmidt (GSA) spectral sharpening, for example, is less affected than an *à trous* wavelet transform (ATWT) (Baronti et al. 2011; Aiazzi et al. 2012). For approaches that rely on spatial filtering, the choice of filter is crucial and could cause severe spatial artifacts. Good results depend on the proper choice of the kernel and filtering method. In general, wavelet- and HPF-based methods in the context of pansharpening are more sensitive to misregistration than Brovey transform, Gram-Schmidt (GS), adaptive generalized intensity hue saturation (GIHS) transform, or the algorithm of University of New Brunswick, that is, UNB Pansharp (Xu et al. 2014).

The registration process is necessary due to the different viewpoints of multi-temporal acquisitions of the same sensor or multi-sensor image acquisitions. Image registration methods are grouped into two categories, namely, area-based and feature-based methods. Area-based approaches follow the matching of the image intensity of corresponding regions. Feature-based techniques match corresponding objects, such as points (road intersections, region corners), lines (roads, boundaries, coastlines), or regions (forests, lakes, fields) (Dawn et al. 2010). In multi-sensor registration, it is important that the various features are detectable in both data types. Originally, the registration process was a manual endeavor, in which the user identified corresponding points in the two images by visual inspection. Nowadays, this process is mostly automated. Owing to the variations in intensity in multi-temporal and multi-sensor remote sensing data, feature detection and matching is more popular (Zhang, Q. et al. 2014). The complexity in the process is the matching of features in diverse data sets, for example, matching optical and radar imagery by identifying common elements. But even in multi-temporal images of the same sensors, corresponding features can appear different due to different imaging conditions (Zitová and Flusser 2003). After feature matching, the registration process requires the estimation of a transformation model that can be more or less complex. An example is the scale-invariant feature transform (SIFT). In the case of remote sensing, it is possible to use orbit ephemeris data to position each acquired pixel in space and predict its location on Earth. Including the terrain model and the geometric viewing parameters of a particular sensor, an accurate reconstruction and geometrically corrected image can be produced. Image registration and geocoding include the resampling of image data to a common pixel spacing and map projection, the latter only in the case of geocoding. Modern systems unify the resampling process for geocoding and fusion in one process so that the data are only resampled once to avoid information loss and the introduction of spatial artifacts.

Another important fact to be considered in pixel-based fusion is the effect of changes between the different acquisition dates of the input images.

Obviously, a combination of changed objects into one fused image can cause spectral artifacts if these changes are not considered or if RSIF is used in an uncontrolled manner. In change detection, the detection of changed objects using RSIF is an explicit goal. In other cases, for example, pansharpening changed objects cause temporal misalignment. Changes could be as simple as seasonal effects on vegetation or water levels. Some RSIF methods exhibit stronger distortions than others (Aiazzi et al. 2011). For example, ATWT can preserve multispectral content better than GSA (Aiazzi et al. 2012). In any case, it requires a conscious selection of data and processes for successful implementation.

Further details on preprocessing, techniques, accuracy assessment, and applications of pixel-based fusion can be found in the successive chapters since the book is meant to guide through the processes necessary for image fusion in remote sensing.

2.5 Feature Level

Complex object structures require more complex processing. Land cover/ land use mapping studies often rely on remote sensing feature-based fusion for classification to obtain a better and more complete description of the objects on the ground. In particular, complex urban areas require additional input into the fusion process because of the man-made variability in urban objects based on their inherent spectral heterogeneity (Powell et al. 2007; Nielsen 2015). *A priori* feature extraction in terms of the urban extent can help the mapping, especially if multiresolution data sets are involved (Gamba et al. 2011; Salentinig and Gamba 2015). Feature fusion implies that the images are processed to extract features of interest using feature extraction methods. Feature-level fusion uses a group of image pixels to form contiguous regions and requires extraction of different features from source data. According to Guyon et al. (2006), "… feature extraction addresses the problem of finding the most compact and informative set of features to improve the efficiency or data storage and processing." Features can be pixel intensities, edges, or texture features. The features involve the extraction of feature primitives like shape, size, or contrast from different types of images of the same geographic location. Different images require different feature extraction techniques to represent the content in terms of texture, shapes, and color distribution. The numerous features can be categorized into two types of features: low level and high level. Low-level features are defined as basic features that can be extracted automatically from an image without any shape information (Nixon 2008). High-level feature extraction accounts for finding shapes in images. Shape extraction involves finding the position, orientation, and size of the targeted shape. The shape can be said to exist and detected as long as there is contrast between the shape and its background.

Feature extraction depends on the type of features to be identified. Spectral features are described through mean and standard deviation. They contain the reflection of the observed ground objects components. Texture is derived through the gray-level co-occurrence matrix (GLCM) and describes the spatial distribution of tonal variations. SIFT delivers structural features that assist in the interpretation and definition of the observed scene. The outcome is a set of complementary feature vectors that can be fused for a final classification of the image (Yanfei et al. 2015).

GLCM successfully describes texture elements in an image (Yanfei et al. 2015). In practice, commonly, five so-called Haralick's feature statistics are used to represent these elements, namely, *energy, entropy, contrast, variance,* and *correlation* (Haralick et al. 1973). Energy is a measure for textural uniformity. Large entropy means that the image is texturally not uniform. It is inversely correlated to energy. Contrast is related to the difference between neighboring pixel values. It is closely correlated to spatial frequency. Variance is a good measure for heterogeneity and increases in the case of pixels that differ from their mean. Last but not least, correlation describes the relationship between digital numbers of pixel pairs. A high correlation represents a linear relationship (Baraldi and Parmiggiani 1995).

The normalization of feature values within a predefined range could be useful in feature-level fusion of remote sensing data. After feature extraction, the values of the identified features in the various images might cover entirely different ranges. The normalization process adjusts minimum and maximum values to allow for an improved cross-sensor comparison and object-based classification (Zhang et al. 2015). Object-based classification follows segmentation, object metrics calculation, and classification using the metrics, that is, shape, texture, etc. (Man et al. 2015).

Feature-level fusion forms part of object-based image analysis (OBIA). In general, it can be stated that a more complex environment requires higher processing levels and more sophisticated algorithms. This is particularly valid with increasing spatial resolution of the remote sensing data as well as for complex scenes observed, that is, urban areas. Therefore feature-level fusion methods, such as layer-stacking and ensemble-learning methods, are applied in land use/land cover (LULC) classification (Zhang, Y. et al. 2014). Especially in an urban context where high-resolution data are a prerequisite and land cover is complex in nature, feature-level fusion is of interest. A classification process that depends on information provided at pixel level will fail in an urban environment on a large scale. Here, OBIA (feature level) approaches tend to be much more successful since they rely on segments of homogeneous pixels of similar radiometry. Feature-level fusion allows the integration of other data types apart from images. An example for feature fusion is the combination of LiDAR digital elevation and surface model data with imagery. A potential flowchart for such applications is shown in

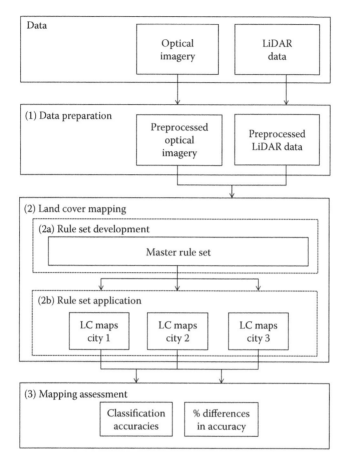

FIGURE 2.7
Flowchart for feature-level fusion integrating high-resolution optical images with LiDAR data. (Adapted from Berger, C. et al. 2013. *IEEE Journal of Selected Topics in Applied Earth Observations and Remote Sensing* 6(5): 2196–2211.)

Figure 2.7. In practice, this approach has already shown its great potential for accurate and robust urban land cover mapping (Berger et al. 2013). The quality of the output largely depends on the proper selection of input data, the segmentation process for suitable features, and the fusion process itself. The combination of hyperspectral data with LiDAR in combination with an attribute profile (AP) advanced classification on feature basis provides a solution toward fully automated procedures. The feature extraction process on the LiDAR data is automated and dimensionality reduction for the hyperspectral data is achieved by PCA. The fusion is carried out as a stacked vector prior to classification (Ghamisi et al. 2015).

Developments in the research on suitable OBIA techniques move toward object identification based on spectral or cumulative frequency distribution.

This improves the performance of the classifier since the derived features are of higher quality based on curve matching rather than using traditional statistical parameters (Zhou and Qiu 2015). LiDAR has the advantage of introducing a third dimension, which facilitates the distinction of different features.

2.6 Decision Level

Fusion at the decision or the so-called symbol level takes place after the images have been processed for each modality separately. The extracted "useful" information is then fused to obtain a final decision. The outcome depends very much on the design of the decision rules. Decision fusion is defined as "the highest level of information fusion, which reinforces the common interpretation, resolves differences" while providing "a better understanding of the observed objects" (Du et al. 2013).

Decision-level fusion approaches are Bayesian inference, Dempster–Shafer theory (DST), artificial neural networks, weighting methods, abductive reasoning, and semantic methods. Inference is a form of pattern matching. The Bayesian method combines evidence related to the probability theory rules. It requires the knowledge of *a priori* probabilities. The Dempster–Shafer inference forms a generalization of the Bayesian method. It helps to represent incomplete knowledge and can describe uncertainty. It allows the weighting of different information input. DST estimates probability mass functions to assign classes to features extracted from multi-sensor data. It comprises reasoning, weight, and probability evidence (Saeidi et al. 2014). The abductive reasoning approach searches for the best probability of the information to be true. It follows a reasoning pattern and can involve neural networks or fuzzy logic. Semantic methods utilize semantic data of different origin to derive the interpretation. It requires a known set of knowledge, which can be used to match the information (Castanedo 2013).

In decision-level fusion, the source data can be very diverse. However, the data/information needs a common reference to establish a proper relationship. In the context of spatial information, this is the common geographic reference. Different layers of spatial information can be consulted to conclude to a final decision and therefore classification of the data. Therefore, decision fusion requires the processing from signal to information before the fusion process takes place. All input data are processed separately with data-specific procedures and algorithms to arrive at the desired information. An illustrative example is the study of different tropical forest vegetation from Landsat TM combined with knowledge and other geographical data. A rule-based model is implemented to identify different forest types from remote sensing imagery and GIS layers. The GIS layers consist of a digital elevation model, rainfall, and temperature (Yang and Huang 2015). The data flow is illustrated in Figure 2.8. The parameters are named by the following abbreviations:

- NDVIs for the years 1997 and 2000: NDVI97 and NDVI2000
- Average precipitation: PA
- Accumulative temperature above 10°C: T10
- The NDVI difference between 1997 and 2000: NDVI32

In the context of remote sensing image classification, the combination of different classifier outputs follows decision-level fusion. These techniques are termed *stacking, ensemble classification,* or *meta-learning* in the literature and belong to the field of data mining. The choice of decision-level fusion makes sense because it leads to higher precision of the output map. In this case, the fusion process does not combine multimodal data but the results of different processing methods. Remote sensing data are classified using a set of classifiers resulting in a number of maps. The second step comprises the actual fusion process, applying a meta-classifier to produce the final fused map as illustrated in Figure 2.9. The quality increases with the appropriate selection of base classifiers. The decision is taken by a meta-classifier that is run with different training data or voting algorithms (Clinton et al. 2015).

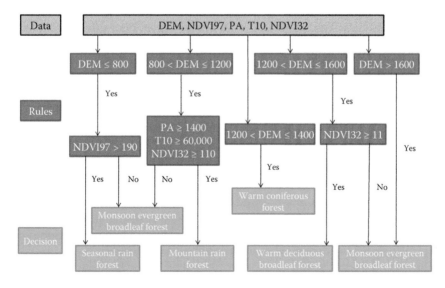

FIGURE 2.8
Example of decision-level fusion to determine different tropical forest vegetation types. (Adapted from Yang, C. and H. Huang. 2015. *Natural Hazards*: 1–11. http://link.springer.com/article/10.1007/s11069-015-1919-z)

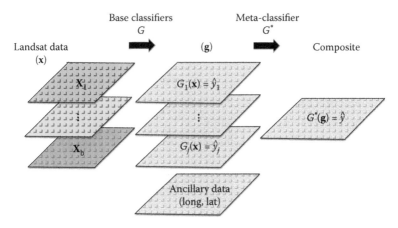

FIGURE 2.9
Geographic stacking using decision-level fusion; (**x**) forms the image with b number of bands, (**g**) is the vector of base classifier (G) guesses, (G^*) is the meta-classifier, which produces the composite. (Adapted from Clinton, N. et al. 2015. *ISPRS Journal of Photogrammetry and Remote Sensing* 103: 57–65.)

2.7 Hybrid-Level Fusion

The most accurate, flexible, and adaptive implementation of fusion is the use of multiple levels. The implementation of various levels simultaneously allows the consideration of advantages of the various levels while omitting the disadvantages. The result is a smart but complex decision support system. Since it exceeds the context of this book, we would like to illustrate the concept with three examples of hybrid-level fusion in this section. The combination of pixel, feature, and/or decision-level fusion maximizes the quality of the fused product and benefits from different RSIF techniques with their individual advantages, that means an improvement of fusion algorithm performance.

The first example describes the optimization of a pansharpening process. It uses a context-adaptive wavelet-based pansharpening method and its generic version. The decision whether the context-based or generic fusion algorithm is implemented is taken on the basis of object size (scale) in order to produce a sharp image with minimum noise. The preprocessing that has to be done in order to implement the decision level in the entire workflow determines the local scale of a structure by scale-adaptive segmentation. The outcome defines the decision fusion criterion (Bin et al. 2013).

The second example also combines pixel-based fusion (pansharpening for change detection) with decision fusion (produce change map). The result is a sequential fusion scheme to increase the accuracy and reduce errors for the output. The experiment involved five pansharpening methods, namely, generalized intensity hue saturation transform, Gram–Schmidt transform,

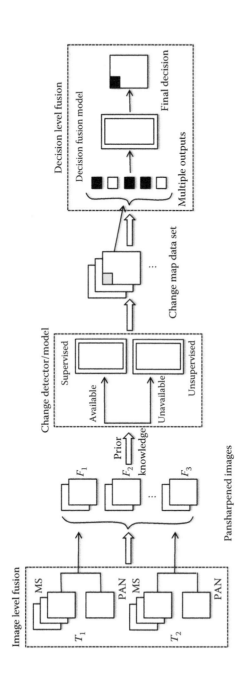

FIGURE 2.10
Procedure of hybrid pixel–decision-level fusion for change detection. (Adapted from Du, P. et al. 2013. *Information Fusion* 14(1): 19–27.)

principal component analysis, high-pass filter fusion, and wavelet transform, which will all be explained in Chapter 4. For the decision level, three different strategies were implemented. The process includes pansharpening, multi-temporal coregistration for change detection, and calculation of change vectors followed by an automated thresholding to generate change maps. Then the final decision is taken using the different change maps as input. The procedure is displayed in Figure 2.10.

Implementing the decision fusion level in this procedure has the advantage that the structure and shape of changed objects are of higher quality in terms of accuracy and completeness. Different pansharpening algorithms (at pixel level) result in different qualities of change maps, which, of course, should be expected. However, with the decision fusion after pixel fusion all algorithms can be considered and an optimized change map can be produced.

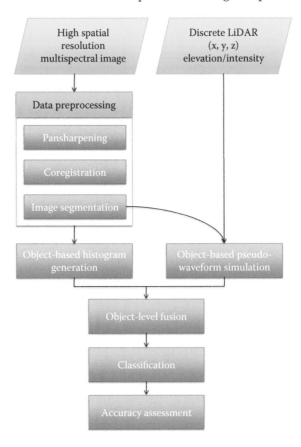

FIGURE 2.11

Flowchart example no. 3 of hybrid pixel–feature-level fusion for urban high-resolution land use mapping. (Adapted from Zhou, Y. and F. Qiu. 2015. *ISPRS Journal of Photogrammetry and Remote Sensing*, 101: 221–232.)

The specific example showed that pansharpening with decision fusion leads to an increased overall accuracy. The decision level reduces omission and commission errors of the change detection results (Du et al. 2013).

The third case we would like to discuss is the combination of pixel and feature-level fusion for urban land use classification. The process is illustrated as a flowchart in Figure 2.11. High-resolution multispectral imagery is fused with 3D LiDAR data including its intensity information. First, the WorldView-2 image is pansharpened (pixel-level fusion) and then segmented to extract the features. From discrete-return LiDAR pseudo-waveforms are synthesized. Finally, the LiDAR pseudo-waveforms are fused with the spectral histograms of the WorldView-2 imagery (feature-level fusion) prior to running the classification process, resulting in the final land use map.

Apparently the fused data significantly improved the OBIA classification, increasing the overall classification accuracy from 89.97% to 97.58% (Zhou and Qiu 2015).

2.8 Summary

Chapter 2 started with the various definitions of image and data fusion used in the literature. We then showed the position of pixel-based image fusion levels within the overall context of data fusion. The chapter describes and explains the four levels of fusion, namely, subpixel, pixel, feature, and decision-based levels. We discussed how and when which level of fusion is most appropriate. Although the focus of this book is on remote sensing image fusion, that is, pixel-based methods, this chapter also contains a section on the value of image and data fusion using remote sensing imagery, with other data such as information from ground truth observations, maps, Internet resources, a geographic information system, and other types of data, leading to higher levels of fusion in remote sensing. Before actually giving and discussing all the various remote sensing image fusion algorithms in use today in Chapter 4, we describe the many preprocessing steps that need to be carried out prior to fusing different images in the next chapter.

References

Aiazzi, B., L. Alparone, S. Baronti et al. 2011. Effects of multitemporal scene changes on pansharpening fusion. Paper read at *2011 6th International Workshop on the Analysis of Multi-Temporal Remote Sensing Images (Multi-Temp)*, July 12–14, 2011.

Aiazzi, B., L. Alparone, S. Baronti, A. Garzelli, and M. Selva. 2012. Twenty-five years of pansharpening: A critical review and new developments. In *Signal and Image Processing for Remote Sensing*, edited by C. H. Chen. Boca Raton, FL: CRC Press, pp. 533–548.

Amoros-Lopez, J., L. Gomez-Chova, L. Alonso, L. Guanter, J. Moreno, and G. Camps-Valls. 2011. Regularized multiresolution spatial unmixing for ENVISAT/MERIS and landsat/TM image fusion. *IEEE Geoscience and Remote Sensing Letters* 8 (5):844–848.

Baraldi, A. and F. Parmiggiani. 1995. An investigation of the textural characteristics associated with gray level cooccurrence matrix statistical parameters. *IEEE Transactions on Geoscience and Remote Sensing* 33 (2):293–304.

Baronti, S., B. Aiazzi, M. Selva, A. Garzelli, and L. Alparone. 2011. A theoretical analysis of the effects of aliasing and misregistration on pansharpened imagery. *IEEE Journal of Selected Topics in Signal Processing* 5 (3):446–453.

Berger, C., M. Voltersen, S. Hese, I. Walde, and C. Schmullius. 2013. Robust extraction of urban land cover information from HSR multi-spectral and LiDAR data. *IEEE Journal of Selected Topics in Applied Earth Observations and Remote Sensing,* 6 (5):2196–2211.

Bin, L., M. M. Khan, T. Bienvenu, J. Chanussot, and Z. Liangpei. 2013. Decision-based fusion for pansharpening of remote sensing images. *IEEE Geoscience and Remote Sensing Letters* 10 (1):19–23.

Castanedo, F. 2013. A review of data fusion techniques. *The Scientific World Journal* 2013:19.

Clinton, N., L. Yu, and P. Gong. 2015. Geographic stacking: Decision fusion to increase global land cover map accuracy. *ISPRS Journal of Photogrammetry and Remote Sensing* 103:57–65.

Dawn, S., V. Saxena, and B. Sharma. 2010. Remote sensing image registration techniques: A survey. In *Image and Signal Processing*, edited by A. Elmoataz, O. Lezoray, F. Nouboud, D. Mammass, and J. Meunier. Berlin, Heidelberg: Springer, pp. 103–112.

Delalieux, S., P. J. Zarco-Tejada, L. Tits, M. A. J. Bello, D. S. Intrigliolo, and B. Somers. 2014. Unmixing-based fusion of hyperspatial and hyperspectral airborne imagery for early detection of vegetation stress. *IEEE Journal of Selected Topics in Applied Earth Observations and Remote Sensing* 7 (6):2571–2582.

Du, P., S. Liu, J. Xia, and Y. Zhao. 2013. Information fusion techniques for change detection from multi-temporal remote sensing images. *Information Fusion* 14 (1):19–27.

Elmenreich, W. 2002. Sensor fusion in time-triggered systems. Dissertation, Faculty of Technical Natural Sciences and Informatics, Technical University Vienna, Vienna, Austria.

Farah, I. R. and M. B. Ahmed. 2010. Towards an intelligent multi-sensor satellite image analysis based on blind source separation using multi-source image fusion. *International Journal of Remote Sensing* 31 (1):13–38.

Gamba, P., M. Aldrighi, and M. Stasolla. 2011. Robust extraction of urban area extents in HR and VHR SAR images. *IEEE Journal of Selected Topics in Applied Earth Observations and Remote Sensing* 4 (1):27–34.

Ghamisi, P., J. A. Benediktsson, and S. Phinn. 2015. Land-cover classification using both hyperspectral and LiDAR data. *International Journal of Image and Data Fusion* 6 (3):189–215.

Gomez-Chova, L., D. Tuia, G. Moser, and G. Camps-Valls. 2015. Multimodal classification of remote sensing images: A review and future directions. *Proceedings of the IEEE* 103 (9):1560–1584.

Guyon, I., S. Gunn, M. Nikravesh, and L. Zadeh (eds.). 2006. Feature extraction. In *Foundations and Applications*. Berlin: Springer Science & Business Media, Vol. 207, 778 pp.

Haralick, R. M., K. Shanmugam, and I. H. Dinstein. 1973. Textural features for image classification. *IEEE Transactions on Systems, Man and Cybernetics SMC-3* (6):610–621.

MacDonald, R. 1997. From space data to information. Paper read at *Joint ISPRS Working Groups I/1, I/3, IV/4 Workshop on Sensors and Mapping from Space*, September 29–October 2, 1997, Hannover, Germany.

Man, Q., P. Dong, and H. Guo. 2015. Pixel- and feature-level fusion of hyperspectral and LiDAR data for urban land-use classification. *International Journal of Remote Sensing* 36 (6):1618–1644.

McKendall, R. and M. Mintz. 1988. Robust fusion of location information. Paper read at *Proceedings of the 1988 IEEE International Conference on Robotics and Automation, 1988*, April 24–29, 1988.

Mitchell, H. B. 2010. *Image Fusion: Theories, Techniques and Applications*. Heidelberg: Springer.

Nielsen, M. M. 2015. Remote sensing for urban planning and management: The use of window-independent context segmentation to extract urban features in Stockholm. *Computers, Environment and Urban Systems* 52:1–9.

Nixon, M. 2008. *Feature Extraction & Image Processing*. 2nd ed. London: Academic Press.

Paradis, S., B. A. Chalmers, R. Carling, and P. Bergeron. 1997. Toward a generic model for situation and threat assessment. Paper read at *Digitization of the Battlefield II*, April 21, 1997, at Orlando, FL, USA, pp. 171–182.

Pohl, C. and J. L. van Genderen. 1998. Review article multisensor image fusion in remote sensing: Concepts, methods and applications. *International Journal of Remote Sensing* 19 (5):823–854.

Powell, R. L., D. A. Roberts, P. E. Dennison, and L. L. Hess. 2007. Sub-pixel mapping of urban land cover using multiple endmember spectral mixture analysis: Manaus, Brazil. *Remote Sensing of Environment* 106 (2):253–267.

Qunming, W., P. M. Atkinson, and S. Wenzhong. 2015. Fast subpixel mapping algorithms for subpixel resolution change detection. *IEEE Transactions on Geoscience and Remote Sensing* 53 (4):1692–1706.

Saeidi, V., B. Pradhan, M. O. Idrees, and Z. Abd Latif. 2014. Fusion of airborne LiDAR with multispectral SPOT 5 image for enhancement of feature extraction using Dempster-Shafer theory. *IEEE Transactions on Geoscience and Remote Sensing* 52 (10):6017–6025.

Salentinig, A. and P. Gamba. 2015. Combining SAR-based and multispectral-based extractions to map urban areas at multiple spatial resolutions. *IEEE Geoscience and Remote Sensing Magazine* 3 (3):100–112.

Wald, L. 1998. A European proposal for terms of reference in data fusion. Paper read at *Resource and Environmental Monitoring*, Budapest, Hungary.

Waltz, E. L. and J. Llinas. 1990. *Multisensor Data Fusion*. Norwood, MA: Artech House Publishers.

White Jr, F. E. (eds.) 1987. Data fusion lexicon. In *Joint Directors of Laboratories, Technical Panel for C3*. San Diego, CA: JDL Data Fusion Subgroup, 13 pp.

Xu, Q., Y. Zhang, and B. Li. 2014. Recent advances in pansharpening and key problems in applications. *International Journal of Image and Data Fusion* 5 (3):175–195.

Yanfei, Z., Z. Qiqi, and Z. Liangpei. 2015. Scene classification based on the multifeature fusion probabilistic topic model for high spatial resolution remote sensing imagery. *IEEE Transactions on Geoscience and Remote Sensing* 53 (11):6207–6222.

Yang, C. and H. Huang. 2015. Mapping tropical forest vegetation from Landsat TM images based on fusion of knowledge and geo-data. *Natural Hazards*: 1–11. http://link.springer.com/article/10.1007/s11069-015-1919-z.

Zhang, H., H. Lin, and Y. Li. 2015. Impacts of feature normalization on optical and SAR data fusion for land use/land cover classification. *IEEE Geoscience and Remote Sensing Letters* 12 (5):1061–1065.

Zhang, Q., Z. Cao, Z. Hu, Y. Jia, and X. Wu. 2014. Joint image registration and fusion for panchromatic and multispectral images. *IEEE Geoscience and Remote Sensing Letters* 12 (3):467–471.

Zhang, Y., H. Zhang, and H. Lin. 2014. Improving the impervious surface estimation with combined use of optical and SAR remote sensing images. *Remote Sensing of Environment* 141:155–167.

Zhou, Y. and F. Qiu. 2015. Fusion of high spatial resolution WorldView-2 imagery and LiDAR pseudo-waveform for object-based image analysis. *ISPRS Journal of Photogrammetry and Remote Sensing* 101:221–232.

Zitová, B. and J. Flusser. 2003. Image registration methods: A survey. *Image and Vision Computing* 21 (11):977–1000.

3

Preprocessing

Chapter 3 provides the reader with an insight to intelligent data selection for the fusion process followed by considerations for preprocessing the images. Prior to the fusion process, the images need to be corrected for platform-, sensor-, and environment-induced distortions. These corrections are image specific and therefore relevant for high-quality fusion. The last part of this chapter describes optional image enhancement procedures that can help to highlight features of interest in the fused image.

3.1 Data Selection

The choice of input data for the fusion process highly depends on the purpose of image fusion. Data that are useful in one case may be useless in another. The selection very much depends on the observed phenomena, characteristics of the sensors, quality and availability of the data, as well as the accessibility of effective algorithms for information extraction (Salentinig and Gamba 2015). Chapter 6 contains a table referring to the most suitable type of data to assist the selection process (refer to Table 6.3). The availability of suitable remote sensing images is an important criterion even though it becomes less and less a problem because of the many satellites that have been and are being launched. This becomes obvious in particular when working with optical remote sensing where cloud cover can be a major hindering factor in obtaining suitable scenes. The use of SAR images, originating from an active microwave sensor delivers data around the clock and independent of cloud cover. Therefore, the choice of VIR images in combination with SAR can become an important option. It works for "cloud removal" as well as updating existing information. The purpose can be related to the desired application where the original input images need to contain the information that we look for. In another context, the combination of multi-temporal image acquisitions from the same or multiple sensors could be of interest. Change detection is a major application in remote sensing, which relies more and more on multi-sensor images due to the variety of sensors available in space. Regular monitoring needs proper image selection to acquire the relevant information. For most applications, users need high spatial and multi-spectral resolution, which is why the combination of high spatial resolution panchromatic with lower-resolution multispectral images has become so popular. This can be extended to high spatial and temporal resolution image fusion, which gains momentum in the remote sensing society. High spatial resolution image acquisition requires more time to cover Earth's surface than lower spatial resolution sensors that have a larger swath and a more frequent revisit time. Spatial and temporal image fusion (STF) plays an important role in environmental monitoring and change detection.

Some fusion techniques are limited to a certain number of bands. Researchers have found three solutions to work around this problem:

1. Select the most appropriate bands
2. Modify the algorithm to suit more bands
3. Repeat the algorithm until all bands have been accommodated

Early implementations went for solution number 1. Some techniques used to be limited in the number of input bands that can be fused (e.g., IHS), while others can be performed with any number of selected input bands. There is a method that relies on statistics in order to select the data containing most of the variance. This is the selection method developed by Chavez et al. (1982) called *optimum index factor* (OIF) mathematically described in Equation 3.1:

$$\text{OIF} = \frac{\sum_{i=1}^{3} \sigma_i}{\sum_{j=1}^{3} |cc_j|} \tag{3.1}$$

where σ_i is the standard deviation of digital numbers for band i and cc_j is the correlation coefficient between any two of the three bands. If more than three bands are involved, 3 can be replaced by any other number of bands.

Another approach is to select the bands, which are the most suitable to a certain application. This requires *a priori* knowledge by the user (Sheffield 1985). Kaufmann and Buchroithner (1994) suggest selecting the three bands with the highest variance. The principal component analysis is another solution to reduce the number of channels containing the majority of the variance for the purpose of image fusion (Yésou et al. 1993; Licciardi et al. 2012). Others provide a certain band combination based on an algorithm which takes into account the statistics of the scene, including correlations between image channels (Sheffield 1985).

In the context of hyperspectral data, the band selection receives an even more important role. Owing to the vast amount of overlapping bands, the selection of appropriate bands becomes vital. The two approaches to reduce dimensionality rely on band selection and feature extraction (Kaewpijit et al. 2003). Algorithms used in this context are PCA and wavelet decomposition.

In hyperspectral image processing, band selection has to be automated due to the large number of bands concerned. It is crucial not to lose any information in this process. Researchers developed solutions that are based on a band correlation minimization approach (Chein and Su 2006).

It should not be underestimated that often financial constraints hinder the optimum selection of data. So the selection process is a compromise between potential and reality. Another factor is the access to suitable software to process a particular data type. A natural limitation is the availability of suitable data on a particular area, which could be hindered by cloud cover or acquisition time. The financial aspect as well as the acquisition period is gradually being reduced since the obtainability of free data, such as the Landsat (NASA) and Sentinel (ESA) programs. These open data archives will boost RSIF and remote sensing applications in general. Often, open source software and the provision of software tools with the data set facilitate the exploitation. An example is the Sentinel Toolbox.

3.2 General Workflow

Image fusion takes place at pixel level. It requires sensor-specifically corrected data and alignment. Influences that affect and falsify the values collected by a certain sensor have to be removed prior to image fusion. In addition, the multi-sensor images have to be coregistered and geometrically corrected to ensure that the data coincide on a pixel-by-pixel basis and refers to the same location on the ground. For the integration in a GIS or the Digital Earth environment, a geocoding is indispensable. Figure 3.1 illustrates the workflow for image fusion. The actual information extraction takes place after the fusion process. If this is not the case, the process is called feature or decision fusion as described in Chapter 2.

Depending on the type of sensor and platform, data acquired by remote sensing techniques contain a number of geometric and radiometric distortions. For the integrative processing and interpretation of multi-temporal or multi-sensor image data, the rectification and restoration of the acquired images is a prerequisite. The degree of processing and correction depends on the images to be evaluated and on the intended application. In some cases, it may be sufficient to correct only for systematic errors and to shift the images to coregister them. In others, the images have to be corrected and resampled to a common map projection (geocoding). The overall correction of satellite images involves the initial processing of raw image data to eliminate geometric distortions, to calibrate the data radiometrically, and to reduce noise present in the data. The following sections contain information about the individual requirements that need to be fulfilled preceding RSIF.

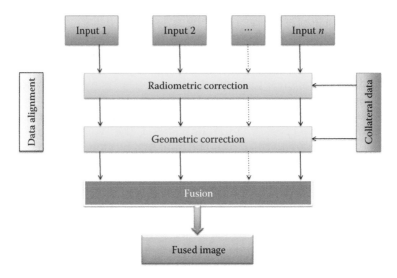

FIGURE 3.1
General workflow in RSIF.

3.3 Sensor-Specific Corrections

The radiometric corrections necessary to remove variations in the radiometry that influence the compatibility of multi-temporal/multi-sensor data are briefly discussed in this section. In RSIF, it is indispensable to provide sensor type-dependent corrected images for the actual fusion process. Any distorted or erroneous data content will falsify the fused result. Owing to the disparate nature of VIR and SAR, we distinguish the two sensor types. Naturally, sensor-specific corrections are also necessary for other kinds of data, such as LiDAR or thermal data. The following sections describe the general correction process needed to successfully fuse optical and radar data.

3.3.1 Radiometric Distortions in VIR Data and Their Corrections

Radiometric distortions in optical satellite images are caused by

1. The atmosphere through which the signal has to pass
2. The sensor that records the data

The sources of radiometric distortions in VIR data are the atmosphere (atmospheric scattering and absorption = attenuation), sensor (striping), and solar illumination. The radiation received at a VIR sensor is altered

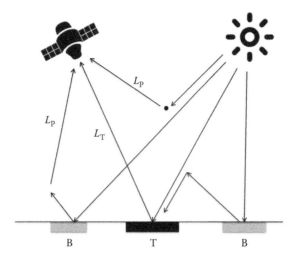

FIGURE 3.2
Components of surface reflected radiation ($L_P + L_T$) received at the remote sensor (T = target, B = background).

compared to the surface spectral reflectance and composed of three different contributions: (1) target (T), (2) background (B), and (3) atmosphere (L) as depicted in Figure 3.2. The atmosphere has severe effects on the signal and cannot be neglected if the image information content of interest is target radiance to provide spectral signatures for objects of interest on Earth's surface. The atmosphere influences the spatial and spectral distribution of the radiation, attenuates target radiation, and scatters the signal due to the particles contained within. Atmospheric correction can be applied absolutely or relatively. Absolute radiometric correction turns the recorded digital numbers into scaled surface reflectance values using actual physical parameters to model atmospheric influence. To do this, atmospheric properties have to be available at time of acquisition. Alternatively, a relative radiometric correction can be carried out using a normalization process. For single images, the histogram is adjusted; multiple images of the same sensor can be normalized using regression.

Normally, the effects induced by the sensor can be neglected due to the larger influence of atmospheric effects on the radiometry. However, we can still experience sensor malfunction, which has to be taken care of. The detectors introduce the most relevant of these distortions. Examples are MODIS where the 15 detectors of band 6 are ineffective and Landsat ETM+ for which the scan line corrector failed (Shen et al. 2015). The resulting images can contain striping effects, which disturb the radiometric integrity of the data. The correction takes into account the cyclic characteristics of the scan line noise. Each pixel is adjusted based on a comparison of the local mean of the line to an unweighted local mean of those closest neighboring lines. However,

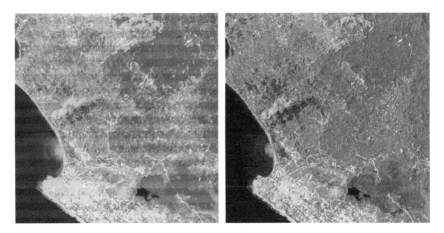

FIGURE 3.3
Correction of sensor-induced striping in TM scenes.

it should be noted that this *de-striping* is a cosmetic act that should be performed after the spectral interpretation or processing of the images. To illustrate the effect, Figure 3.3 shows a Landsat TM scene of an area in Indonesia
(Bengkulu, Sumatra) containing a 16-line striping. The original TM (left) and
the corrected data (right) are depicted in the figure. The data were de-striped
using the algorithm mentioned above. This has to be done prior to image
fusion to take into account the radiometric characteristics of the VIR data
alone, which cannot be separated from SAR, for example, in VIR/SAR fused
imagery.

3.3.2 Radiometric Corrections of SAR Images

There are basically four aspects, which influence the radiometric content of
the SAR image to be discussed in this section. First, there is the so-called
speckle, which reduces the interpretability and effectiveness of target recognition and classification of a SAR image. In polarimetric SAR, speckle does
not only occur in the intensity image of each polarization but also in the
complex cross product terms between the polarizations (Xiaoshuang et al.
2015). A second factor influencing SAR radiometry is the antenna pattern.
Then the section continues with some considerations on the dynamic range
of the data relating to the 16-bit format of SAR data and its conversion to
8-bit. Finally, some aspects of SAR calibration are summarized.

The appearance of SAR images is disturbed by a multiplicative granular
noise pattern, that is, speckle (Gomez et al. 2015). Speckle causes a grainy
appearance (also called *salt and pepper effect*) in SAR images and is an effect of
surface roughness and system factors on image production. It is a phenomenon of coherent scattering and accounts for the interference contribution of
individual scatterers (Schreier 1993). The backscattered signal represented

in one resolution cell (pixel) is composed of contributions from a large number of independent scatterers (e.g., individual leaves from vegetated areas). The interference of these returning waves causes variations in gray levels in adjacent pixels in the image and produces the speckle. Speckle occurs with any active system using coherent waves (e.g., radar, laser). It possesses multiplicative characteristics and limits the radiometric resolution of SAR images significantly. For the interpretation of SAR images, visually as well as digitally, it is often necessary to reduce the speckle (removal is not possible since it is a random effect). This is achieved by averaging. There are various possibilities for speckle reduction: (1) multi-look processing (preprocessing), (2) averaging (pre- or postprocessing), and (3) filtering (postprocessing). All three methods result in a reduced spatial resolution.

3.3.2.1 Multi-Look Processing

Here the movement of the sensor is taken into account, which offers the possibility of receiving a backscattered signal from one target at different positions along the orbit. Instead of using the full Doppler history to generate the best possible resolution image, the history is divided into smaller parts. The available bandwidth in azimuth is divided into several pieces (number of parts = number of looks), which are processed into individual images. They are considered statistically uncorrelated if the bands do not overlap. The average of the separately produced images forms the speckle-reduced lower-resolution multilook SAR image. The achieved speckle reduction is dependent on the number of looks, n, taken. The speckle, along with the spatial resolution, is reduced by a factor of $1/n$. An advantage of the multi-look processing is that the azimuth resolution is adapted to the lower range resolution and square image pixels are produced (Schreier 1993).

3.3.2.2 Spatial Averaging Method

Spatial averaging uses the full-resolution single-look image, which is then low-pass filtered to reduce the high-frequency speckle influence. Another averaging process uses areas of similar ground cover types that are averaged to obtain the expected mean intensity for a homogeneous area. The problem with this method is that additional knowledge about the ground cover and object-based image processing techniques are required.

3.3.2.3 Filtering Techniques

A speckle reduction technique, applied after the actual SAR image has been produced from radar signals, is filtering. Speckle filtering is an active research field while the first speckle-specific filters were developed in the beginning of the 1980s (Jong-Sen 1980; Frost et al. 1982; Kuan et al. 1985). The ideal speckle filter should smooth homogeneous areas but at the same time preserve edges

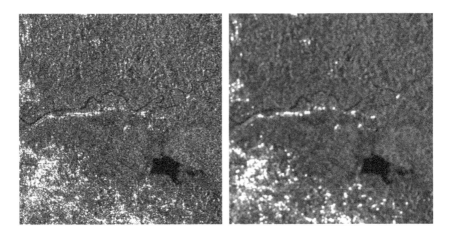

FIGURE 3.4
Original ERS-1 SAR image (left) and its gamma MAP speckle filtered version (right).

and texture, as well as maintain differences between homogeneous areas. In other words, speckle filters aim at the preservation of the mean and the reduction of the standard deviation. Figure 3.4 shows an ERS-1 SAR image that has been speckle filtered. Most of the conventional filters achieve either one aspect or the other. Therefore, speckle-specific ("intelligent") filters had to be developed that act according to local facts based on calculated statistical values. The application of speckle filters is computationally intensive but no longer represents a real constraint due to the existence of powerful hardware (Jong-Sen et al. 2009). Recent developments use more complex concepts, such as the Bayesian nonlocal-means filter (Gomez et al. 2013, 2015). Polarimetric SAR speckle filtering is more complex than it is the case of single-polarization SAR. For polarimetric SAR (PolSAR), the speckle in the complex cross product terms adds to the speckle in the intensity images. Researchers found that it is multiplicative noise along the diagonal terms of the covariance matrix and combine multiplicative and additive noise for off-diagonal terms (Lopez-Martinez and Fabregas 2003). Therefore, PolSAR-specific filters evolved; amongst them are the linear minimum mean-squared error estimator (Jong-Sen et al. 2006), simulated annealing (Schou and Skriver 2001), nonlocal-means (Jiong et al. 2011), and bilateral filtering (D'Hondt et al. 2013).

3.3.2.4 Antenna Pattern

A vertical antenna pattern causes distortions in radar pixel intensities in range direction. Usually, this appears as a gradual increase and then decrease in mean column gray levels. The antenna pattern correction compensates for this nonuniform illumination of the target in range direction. It is based upon least-square polynomials and estimates the mean gray level at each pixel location in the image. In the corrected image, the columns' mean gray levels are equal to the mean gray level of the image. The results of this correction method depend largely on the distribution of cover types in the image. The ideal condition is an even distribution of cover types to ensure that target and signal interaction are a function of range. If this is not the case, the image is divided into subscenes, which contain evenly distributed ground covers.

3.3.2.5 16-Bit to 8-Bit Data Conversion

SAR data are delivered in 16-bit per pixel but some display devices and software packages can only handle 8-bit range data. Additionally, the 16-bit dynamic range of the data requires more storage capacity because of its larger data volume. As a result, the data have to be transformed from 16-bit to 8-bit for further evaluation. The possibilities for converting SAR data from 65,536 possible values to 256 available levels are manifold:

- Square root
- Division by 256
- Nonlinear scaling (histogram equalization)
- Thresholding
- Linear scaling (maximum and minimum, mean and standard deviation, visually assessed scaling)

The first two methods provide quick results but do not take into account how the data are distributed over the full range of values. This can cause a high degree of information loss. The histogram equalization technique provides good results but the conversion is unique to each image and cannot be repeated for different scenes. The thresholding appears to be a quick and simple process, which is repeatable and offers data that can still be calibrated. For this method the distribution of the values is examined in the histogram and the upper limit is determined for display. Next, a constant of division is defined and this divides all values. Any value remaining above 255 is clipped at 255. Likewise, the linear scaling is adaptive to the image data using the minimum and maximum value, the data mean and number of standard deviation, or a visual assessment for the determination of the thresholds.

3.3.2.6 Calibration

In contrast to VIR sensors, the SAR imaging system can be upgraded to a measurement system using a calibration constant which is derived from measurements of known objects (corner reflector). Using this calibration constant K, which depends on the SAR processor used, the image product, and local incidence angle variations, it is possible to calculate the backscatter coefficient (σ_0) of a distributed target from the measured intensity. The goal of SAR image calibration is to create reliable and repeatable measurements of target radar cross section, which implies the possibility of obtaining data with a specific geophysical quantity and of comparing data over similar objects. For many applications, this is a useful processing step in image classification (backscatter models) but it is not a necessity. It allows the combination of products provided by different agencies using different processors, as well as a year-by-year comparison of the information content of the data. It forms a reliable basis for the comparison of data of different objects.

3.3.3 Geometric Corrections

The geometric distortions can be divided into two groups: those that are systematic or predictable, and those that are essentially random, or unpredictable. Systematic distortions are well understood and correctable by applying formulas derived by modeling the sources of the distortions mathematically. Random distortions and residual unknown systematic distortions are corrected by calculating approximating polynomials using well-distributed ground control points (GCPs) occurring in an image.

The geometric distortions of remote sensing data are related to a number of factors, including the following:

- Earth's rotation during image acquisition
- Earth's curvature
- Finite scan rate of some sensors
- Wide field of view of some sensors
- Sensor malfunctions
- Variations in platform altitude, attitude, and velocity
- Panoramic effects related to the imaging geometry
- Ground relief under observation

Another categorization of the sources of distortions results in two groups: acquisition system (observer)-related distortions and object (observed)-related distortions (Earth and atmosphere). The error sources are described in Table 3.1 (Toutin 2011).

Depending on the geometric correction model, the effects are corrected for systematically (sensor model) or generically (polynomial).

3.3.4 Geometric Distortions in VIR Imagery

Optical images are geometrically distorted mainly by the imaging process itself rather than the physical characteristics of the object observed. Still, for some sensors, the shape and movement of Earth play an important role and

TABLE 3.1

Error Sources for Geometric Distortions according to Toutin (2011)

	Advantages	Disadvantages
Acquisition system (observer)	Platform	Variation of movement
		Variation in platform attitude
	Sensor	Variation in sensor mechanics
		Lens distortions
		Viewing angles
		Panoramic effect
	Measuring instrument	Time variations or drift
		Clock synchronicity
Object sensed (observed)	Atmosphere	Refraction and turbulence
	Earth	Curvature
		Rotation
		Topographic effect
	Map	Geoid to ellipsoid
		Ellipsoid to map

have to be taken into account. The sources of distortions that are taken into account in the correction process for optical data are as follows:

- Earth's rotation effects
- Earth's curvature
- Platform variations
- Sensor nonlinearity
- Scan time skew
- Terrain height variations
- Viewing angle

3.3.5 Geometric Effects in SAR Data

Radar sensors, with their side-scanning geometry, are extremely sensitive to variations in terrain height. The same accounts for the shape of the target and its orientation to the radar antenna. The user of SAR data therefore has to deal with foreshortening, layover, and shadow if working in mountainous terrain. These have already been explained in Chapter 1.

As discussed, the location of an object in the image in range direction is determined by the time the signal needs for the two-way travel between the antenna and target. The frequency of the received signal backscattered from a target on Earth's surface depends on the relative velocity of sensor and object. The movement of the sensor platform causes a Doppler frequency shift, that is, the frequency increases and decreases as the sensor–target distance decreases and increases. Since Earth is also moving during the image acquisition, an additional Doppler frequency shift is introduced.

The influence of Earth's rotation and Earth's curvature on the Doppler shift depends on the position of the target within the swath of the system because the radial velocity of a target at near-range is lower than at far-range. Another factor is the geographic latitude of the satellite track (Schreier 1993).

3.4 Geometric Corrections

The purpose of geometric correction is to compensate for the distortions introduced by certain factors so that the corrected image will have the geometric integrity of a map. This is a necessity for many Earth science applications before the information can be extracted, especially if multisource data are simultaneously evaluated in a GIS. The geometric correction is obligatory for the integrative processing and interpretation of multi-temporal or multi-sensor image data. The degree of processing and correction depends on the images to be evaluated and their application.

3.4.1 Terms

The definitions of three common expressions are given here as used by the authors in order to clarify some terms that frequently occur in the following sections. The extent of geometric correction depends on the application. There are three possibilities:

1. Coregistration
2. Geo-referencing
3. Geocoding

In some cases, it is not necessary to rectify data to a certain map projection and assign geographic coordinates. For a simple comparison of two different data sets, it might be sufficient to coregister the images using an affine or polynomial transformation based on tie point measurements. For geo-referencing, the user assigns coordinates to the image pixels via the geometric correction model but does not actually rectify and resample the data. Only the transformation equations are defined in order to be able to relate image processing and interpretation results to a geographic location. The third possibility transforms the image into a map-like object where each pixel is in its geometrically correct position and has coordinates. In the context of pixel-based image fusion, geocoding is an essential preprocessing step, which has to be performed as accurately as possible in order to make the data compatible on a pixel-by-pixel basis.

3.4.2 Geometric Models

There are various possibilities for correcting image data in the multi-sensor environment. Depending on their purpose, images can be registered to other images, maps, or the ground. Two images can be coregistered by registering each of them to a map coordinate base separately, as described below. Alternatively, an image can be registered to a previously geocoded image. If geo-referencing is not important, one image can be chosen as a master, to which the other (known as the slave) is to be registered. In this case, tie points located in two overlapping images are measured to transform one image into the geometry of the other. The images are then comparable on a pixel-by-pixel basis. In rectifying an image to the geometry of a map, the ground control points identified in the image were traditionally determined in the map or by conventional surveying methods. Nowadays, we use Global Navigation Satellite Systems (GNSS), such as GPS (Global Positioning System), GLONASS (GLObalnaja NAvigatsionnaja Sputnikovaja Sistema), Galileo, or the BeiDou Navigation Satellite System. Then, the image will be transformed into the geometry and projection of the map. Relief distortions are not corrected. Third, ground control points and the digital information on the terrain height stored in a DEM are used to rectify the image. In this case, the image is also corrected for local variations in the terrain.

Early work on digital image rectification used geometric models, which were developed in photogrammetry. Nonparametric approaches (e.g., polynomial rectification) as well as parametric methods (differential, projective, and collinearity models) with additional parameters, in combination with least-square adjustments, were suggested for describing the imaging process (Konecny 1979; Novak 1992).

The nonparametric method is based on GCPs and/or tie points (common points in different images), which are used to calculate a polynomial that approximates the image acquisition model. The parametric approach reconstructs the image acquisition procedure and needs the sensor parameters, including the orbit description, to model the image formation. Hereafter, the former method is called the polynomial model, and the latter the sensor model. The use of one or the other depends on the data (observed ground), the availability of ancillary information, and the accessibility of suitable software and hardware.

For RSIF, an accurate geometric processing is indispensable. The input images have to match with great precision so that the data content coincides at pixel level and refers to the same target on the ground. Owing to the disparate nature of geometric influences for different sensors, the geometric processing needs to account for its viewing geometry. The ortho-rectification process itself can be a fusion process, where information on platform, sensor, Earth, and map projection are taken into account. In the following section, we will briefly introduce the different solutions in geometric correction processing. A further in-depth description of issues, models, and processing on the background of fusion can be found in a comprehensive paper by Toutin (2011).

3.4.2.1 Polynomial Rectification

The polynomial method is a relatively simple approach to correct images geometrically. It corrects for distortions of the image relative to a set of control points. It is sensor-independent and based on statistical principles. A number of well-distributed and accurately determined points are selected in both the image and the reference (other image or map as described above). In the correction process, numerous points are located both in the distorted image (column, row numbers) and in the reference map or master image (ground coordinates X, Y, Z). The original image is shifted, rotated, scaled, and warped to fit the reference points.

Different numbers of control points are needed depending on the degree of the polynomial. With increasing order (Equations 3.2 through 3.7), more points are needed to calculate the unknowns. A first-order polynomial describes the translation, rotation, scaling, and obliquity of an image with six unknowns (Equations 3.2 and 3.3).

$$x = a_0 + a_1 x + a_2 y \tag{3.2}$$

$$y = b_0 + b_1 x + b_2 y \tag{3.3}$$

The second-order polynomial adds parameters for torsion and convexity with 12 unknowns (Equations 3.4 and 3.5).

$$x = a_0 + \cdots + a_3 x^2 + a_4 xy + a_5 y^2 \tag{3.4}$$

$$y = b_0 + \cdots + b_3 x^2 + b_4 xy + b_5 y^2 \tag{3.5}$$

The additional parameters in the third order cannot be explained by physical effects anymore and contains 20 unknowns (Equations 3.6 and 3.7) (Toutin 2004).

$$x = a_0 + \cdots + a_6 x^3 + a_7 x^2 y + a_8 xy^2 + a_9 y^3 \tag{3.6}$$

$$y = b_0 + \cdots + b_6 x^3 + b_7 x^2 y + b_8 xy^2 + b_9 y^3 \tag{3.7}$$

Each point delivers at least two measurements (planimetric coordinates: *x, y*). The polynomial equation can be solved after a sufficient number of points have been collected. In the case of redundant information, a least-square adjustment of the measurements is applied to determine the best-fitting polynomial and its accuracy. This method also offers the possibility of detecting blunders.

The polynomial approach corrects the image only locally because it depends on the control point locations and their distribution. For flat terrain and a good distribution of points (no extrapolation), it ensures a well-corrected image. This model is not useful for strong height variations in the terrain and can lead to large errors in regions, which are not covered by control points. It requires a high number of GCPs. The method does not consider sensor characteristics and can therefore be applied to any image data. All the distortions listed above are corrected simultaneously. Much research has gone into the automation of this process. Using feature extraction and matching the image are coregistered using segmentation results (Dawn et al. 2010).

3.4.2.2 Sensor Model

As the name indicates, this model is sensor specific. It requires knowledge of the geometry of image acquisition and orbit parameters, which describe the position and movement of the platform (ephemeris) as well as information on the observed terrain. The elements accounted for are the following:

- Platform (position, velocity, orientation)
- Sensor (orientation angles, IFOV, line, integration time)
- Earth (geoid/ellipsoid including elevation)
- Cartographic projection (ellipsoid, plane)

Software packages for geometric correction are available for many operational remote sensing satellites which deliver data on a continuous basis; these implement the sensor models. The ephemeris data are read from header files of the image data, which usually contain the necessary information. Using a few GCPs and a least-square adjustment can improve the parameters of the sensor models. Recent satellites have a very accurate positioning so that even the additional GCPs are redundant because the orbit description is of very high quality. A digital elevation model provides the necessary information to remove distortions induced by the terrain. The rectification process follows either the direct or the indirect method: the direct method means the input image pixel value is mapped to a calculated position in the output image; the indirect approach interpolates a pixel value from the input image for the selected output image position. Different geometric models are implemented depending on the type of sensor, since different factors affect the image acquisition.

The sensor model, together with the elevation information from a DEM, corrects the image globally and therefore with a higher accuracy than the polynomial method. The disadvantage is that each sensor requires a separate model, information on the sensor and orbit parameters is necessary, and it is computationally more intensive. For images of mountainous areas, a sensor model is inevitable. The implementation of the theory described may deviate in various aspects depending on the software used for the geocoding process.

The developed models are sensor specific. There have been attempts to develop a generic and high-accuracy modeling approach (Cheng et al. 1995). An example is a model developed by Toutin at the Canadian Centre of Remote Sensing (CCRS). It is interesting to know that the model is capable of simultaneously processing various images. More images increase the relative accuracy between them. The approach relies on the above-mentioned collinearity equations, which describe the relationship between the locations of a position in the image and on the ground. The modeling is based on the collinearity conditions, which represent the physical law of the transformation between the image and ground space (Toutin 2004).

3.4.3 Resampling

The gray value for the calculated position in the geometric correction process is assigned using one of the resampling techniques, that is, nearest neighbor, bilinear interpolation, or cubic convolution. Each technique has a difference influence on the final value. Depending on the application, different resampling techniques are chosen. Based on the nature of the different resampling techniques, the digital number of a pixel is changed less or more. Applications that require most original data rely on nearest neighbor; others require a smooth appearance would rely on cubic convolution. The pixels considered identifying the value of the output pixel for nearest neighbor and bilinear interpolation resampling techniques are shown in Figure 3.5. In the figure, the transformed image is shown in red. The original pixels are represented in black. The green arrow indicates the input values that are used to form the new value of the transformed image.

Advantages and disadvantages of the different approaches have to be considered for a final decision on the resampling technique. Table 3.2 lists the decision criteria of the three techniques.

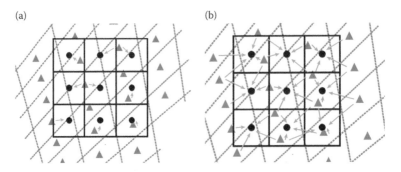

(a)   (b)

FIGURE 3.5
Resampling of remote sensing images: (a) nearest neighbor, (b) bilinear interpolation.

TABLE 3.2

Advantages and Disadvantages of Different Resampling Techniques

	Advantages	**Disadvantages**
Nearest neighbor	• Output values are the original input values • Easy and fast computation	• Choppy, stair-stepped effect • Data values may be lost, while other values may be duplicated
Bilinear interpolation	• Reduced stair-step effect	• Alters original data • Reduces contrast by averaging neighboring pixels • Computationally more expensive
Cubic convolution	• Stair-step effect further reduced • Smoothing of the image	• Strong alteration of original values by averaging a large number of values • Computationally most expensive

Only a few researchers have considered the resampling method in the evaluation of their research on RSIF quality. Apparently the choice of the resampling technique depends on the fusion technique to be applied in order to obtain best possible results, at least for pansharpening. Figure 3.6 compares the results of applying the three resampling types to different fused images (Jawak and Luis 2013). Another factor in choosing the "right"

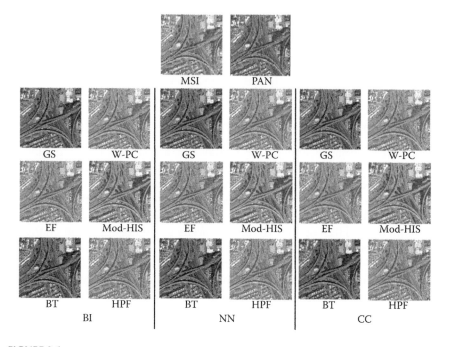

FIGURE 3.6

Resampling effects in different fusion techniques on pansharpening. (Adapted from Jawak, S. D. and A. J. Luis. 2013. *Advances in Remote Sensing* 2 (4): 332–344.)

resampling technique is the intended application. Visual interpretation benefits from cubic convolution because human vision is not disturbed by the stair-stepped effect. However, if automated processes were anticipated, remaining as close as possible to the original value might indicate to choose nearest neighbor for resampling.

3.5 Image Enhancement

The idea of image enhancement is to convert the raw image into something more interpretable for a particular application. The process is supposed to improve visual quality but also increase the potential for automatic information extraction approaches. Another reason for image enhancement is the correction of sensor or environmental condition induced noise. The techniques can be grouped into spatial and frequency domain techniques. For RSIF, individual image enhancement can be a major factor to increase the quality of the fused result, in particular if the quality decreasing factors are sensor specific. An example is the combination of VIR and SAR data where the image content and acquisition process is very different. A general rule of thumb is that the better the image quality that enters the fusion process the better the final fused output. Some researchers consider pansharpening as an image enhancement technique, since it belongs to the group of spatial enhancements. Strictly speaking, it is a fusion process. The following sections briefly discuss fusion-relevant enhancement procedures.

3.5.1 Spatial Filters

Spatial filters enhance images for visual and automated image exploitation. They are also used to extract features. During the filtering process, the digital numbers in the image are altered. Therefore, filtering prior to image fusion should be applied with caution and knowledge about the impact of the filter on the data. Spatial filters can emphasize or suppress different spatial frequencies. High spatial frequency areas show abrupt value changes over a small number of pixels. These features represent field boundaries, coastlines, roads, rivers, etc. A low spatial frequency relates to homogeneous areas where digital numbers only change gradually over a large area. High-pass filters enhance edges and lines (high-frequency areas), while low-pass filters emphasize low-frequency features and suppress high-frequency components and have a smoothing effect. In filtering, the kernel or window size is a key parameter. The larger the kernel the more the original image values are altered and the stronger the effect of the filter.

3.5.2 Frequency Domain

In order to perform image enhancement in the frequency domain, the image can be transformed using a fast Fourier transform (FFT). After enhancement of the data using a filtering algorithm, it is inverse transformed back into the spatial domain. In the frequency domain, brightness, contrast, or the distribution of gray levels can be influenced. Again high- and low-pass filters apply. Interestingly enough, the idea of frequency domain enhancement led to a new fusion algorithm, in which the actual fusion process is done in the frequency domain. This image fusion technique is called Ehlers fusion and will be discussed in Chapter 4.

3.5.3 Histogram Matching

The histogram of two data sets is matched to follow the same distribution of values. Histogram matching results in the same average and standard deviation values of the digital numbers in the image. The matching of the two histograms makes sense where the two images are obtained from different sensors or at different times. Image fusion results improve significantly if the histogram of the replacement image, mostly the high-resolution panchromatic image, is matched to the channel/image to be replaced, for example, the panchromatic image replaces intensity in IHS fusion (Dou and Chen 2008). Matching the histograms minimizes mismatching of brightness during the fusion process and reduces the spectral distortion. The resulting fused image is of higher spectral quality (Amro et al. 2011). The relevance of histogram matching has increased with the extension of panchromatic wavelengths in new optical high-resolution sensors, such as IKONOS, QuickBird, and Landsat 7 in comparison to SPOT or Landsat 4/5. Fusion methods should consider the spectral discrepancy in order to result in high-quality (Xu et al. 2008).

3.6 Summary

In this chapter, we have provided the reader with an overview of the data selection factors that need to be taken into account prior to image fusion. The images need to be corrected for platform, sensor, and environmentally induced distortions. As these distortions are image specific, the need to correct them is critical for a good fused result. The chapter has also shown that the choice of input imagery for the fusion process also depends on the purpose or intended application of the image fusion. As image fusion takes place at the pixel level, there is a need for sensor-specific corrected data prior to fusion. All the various parameters involved in radiometric and geometric correction approaches have been presented and explained. The various image enhancement techniques commonly used, such as the various spatial filtering techniques, as well as filtering in the frequency domain, and histogram matching approaches have been presented.

Having now introduced the objectives of image fusion, explained the various levels of image fusion, and in this chapter the preprocessing steps that are required prior to the fusion process, in the next chapter, we present image fusion algorithms in use today.

References

Amro, I., J. Mateos, M. Vega, R. Molina, and A. Katsaggelos. 2011. A survey of classical methods and new trends in pansharpening of multispectral images. *EURASIP Journal on Advances in Signal Processing* 2011 (1):79.

Chavez, P. S., G. I. Berlin, and I. B. Sowers. 1982. Statistical methods for selecting Landsat MSS ratios. *Journal of Applied Photographic Engineering* 8 (1):23–30.

Chein, I. C. and S. Wang. 2006. Constrained band selection for hyperspectral imagery. *IEEE Transactions on Geoscience and Remote Sensing* 44 (6):1575–1585.

Cheng, P., T. Toutin, and C. Pohl. 1995. A comparison of geometric models for multisource data fusion. *International Symposium on Remote Sensing, GIS and GPS in Sustainable Development and Environmental Monitoring, GeoInformatics*, Hong Kong, 11–17.

Dawn, S., V. Saxena, and B. Sharma. 2010. Remote sensing image registration techniques: A survey. In *Image and Signal Processing*, edited by A. Elmoataz, O. Lezoray, F. Nouboud, D. Mammass, and J. Meunier. Berlin, Heidelberg: Springer.

D'Hondt, O., S. Guillaso, and O. Hellwich. 2013. Iterative bilateral filtering of polarimetric SAR data. *IEEE Journal of Selected Topics in Applied Earth Observations and Remote Sensing* 6 (3):1628–1639.

Dou, W. and Y. Chen. 2008. An improved IHS image fusion method with high spectral fidelity. *The International Archives of the Photogrammetry, Remote Sensing and Spatial Information Sciences* 37:1253–1256.

Frost, V. S., J. A. Stiles, K. S. Shanmugan, and J. Holtzman. 1982. A model for radar images and its application to adaptive digital filtering of multiplicative noise. *IEEE Transactions on Pattern Analysis and Machine Intelligence* PAMI-4 (2):157–166.

Gomez, L., M. E. Buemi, J. C. Jacobo-Berlles, and M. E. Mejail. 2015. A new image quality index for objectively evaluating despeckling filtering in SAR images. *IEEE Journal of Selected Topics in Applied Earth Observations and Remote Sensing* PP (99):1–11.

Gomez, L., C. G. Munteanu, M. E. Buemi, J. C. Jacobo-Berlles, and M. E. Mejail. 2013. Supervised constrained optimization of Bayesian nonlocal means filter with Sigma preselection for despeckling SAR images. *IEEE Transactions on Geoscience and Remote Sensing* 51 (8):4563–4575.

Jawak, S. D. and A. J. Luis. 2013. A comprehensive evaluation of PAN-sharpening algorithms coupled with resampling methods for image synthesis of very high resolution remotely sensed satellite data. *Advances in Remote Sensing* 2 (4): 332–344.

Jiong, C., C. Yilun, A. Wentao, C. Yi, and Y. Jian. 2011. Nonlocal filtering for polarimetric SAR data: A pretest approach. *IEEE Transactions on Geoscience and Remote Sensing* 49 (5):1744–1754.

Jong-Sen, L. 1980. Digital image enhancement and noise filtering by use of local statistics. *IEEE Transactions on Pattern Analysis and Machine Intelligence* PAMI-2 (2):165–168.

Jong-Sen, L., M. R. Grunes, D. L. Schuler, E. Pottier, and L. Ferro-Famil. 2006. Scattering-model-based speckle filtering of polarimetric SAR data. *IEEE Transactions on Geoscience and Remote Sensing* 44 (1):176–187.

Jong-Sen, L., L. Jong-Sen, W. Jen-Hung, T. L. Ainsworth, C. Kun-Shan, and A. J. Chen. 2009. Improved Sigma filter for speckle filtering of SAR imagery. *IEEE Transactions on Geoscience and Remote Sensing* 47 (1):202–213.

Kaewpijit, S., J. Le Moigne, and T. El-Ghazawi. 2003. Automatic reduction of hyperspectral imagery using wavelet spectral analysis. *IEEE Transactions on Geoscience and Remote Sensing* 41 (4):863–871.

Kaufmann, K.-U. and M. F. Buchroithner. 1994. Herstellung und Anwendungsmöglichkeiten von Satellitenbildkarten durch digitale Kombination von Landsat TM und KWR-1000. *Wissenschaftliche Zeitschrift der Technischen Universität Dresden* 43 (5):64–69.

Konecny, G. 1979. Methods and possibilities for digital differential rectification. *Photogrammetric Engineering and Remote Sensing* 45:727–734.

Kuan, D. T., A. A. Sawchuk, T. C. Strand, and P. Chavel. 1985. Adaptive noise smoothing filter for images with signal-dependent noise. *IEEE Transactions on Pattern Analysis and Machine Intelligence* PAMI-7 (2):165–177.

Licciardi, G., M. Khan, J. Chanussot, A. Montanvert, L. Condat, and C. Jutten. 2012. Fusion of hyperspectral and panchromatic images using multiresolution analysis and nonlinear PCA band reduction. *EURASIP Journal on Advances in Signal Processing* 2012 (1):1–17.

Lopez-Martinez, C. and X. Fabregas. 2003. Polarimetric SAR speckle noise model. *IEEE Transactions on Geoscience and Remote Sensing* 41 (10):2232–2242.

Novak, K. 1992. Rectification of digital imagery. *Photogrammetric Engineering and Remote Sensing* 58 (3):339–344.

Salentinig, A. and P. Gamba. 2015. Combining SAR-based and multispectral-based extractions to map urban areas at multiple spatial resolutions. *IEEE Geoscience and Remote Sensing Magazine* 3 (3):100–112.

Schou, J. and H. Skriver. 2001. Restoration of polarimetric SAR images using simulated annealing. *IEEE Transactions on Geoscience and Remote Sensing* 39 (9):2005–2016.

Schreier, G. 1993. *SAR Geocoding: Data and Systems*. Berlin: Wichmann.

Sheffield, C. 1985. Selecting band combinations from multispectral data. *Photogrammetric Engineering and Remote Sensing* 51:681–687.

Shen, H., X. Li, Q. Cheng et al. 2015. Missing information reconstruction of remote sensing data: A technical review. *IEEE Geoscience and Remote Sensing Magazine* 3 (3):61–85.

Toutin, T. 2004. Review article: Geometric processing of remote sensing images: Models, algorithms and methods. *International Journal of Remote Sensing* 25 (10):1893–1924.

Toutin, T. 2011. State-of-the-art of geometric correction of remote sensing data: A data fusion perspective. *International Journal of Image and Data Fusion* 2 (1):3–35.

Xiaoshuang, M., S. Huanfeng, Z. Liangpei, Y. Jie, and Z. Hongyan. 2015. Adaptive anisotropic diffusion method for polarimetric SAR speckle filtering. *IEEE Journal of Selected Topics in Applied Earth Observations and Remote Sensing* 8 (3):1041–1050.

Xu, J., Z. Guan, and J. Liu. 2008. An improved IHS fusion method for merging multispectral and panchromatic images considering sensor spectral response. *The International Archives of the Photogrammetry, Remote Sensing and Spatial Information Sciences* 37:1169–1174.

Yésou, H., Y. Besnus, and J. Rolet. 1993. Extraction of spectral information from Landsat TM data and merger with SPOT panchromatic imagery—A contribution to the study of geological structures. *ISPRS Journal of Photogrammetry and Remote Sensing* 48 (5):23–36.

4

Fusion Techniques

This chapter provides practicable image fusion techniques available for remote sensing images. The number of algorithms has increased over the years, and it has become a challenge to organize and present the various possibilities of fusing remote sensing images. In particular, it has become extremely difficult to identify the various algorithms since they often appear under different names. This is due to the fact that many fusion algorithms have been implemented in commercial software. Different software providers chose different names to supply the same thing. Another difficulty in listing available algorithms occurs in the identification of who really originated the approach because over the years published research on the development of new algorithms has not been cited from the originator. Authors rather cite citations of citations, which make it sometimes impossible to track the material back to its origin. Together with the confusion of multiple names for one algorithm, the challenge to master is identification of the original algorithm along with using its accepted terminology. In this chapter, we have tried to present the basic underlying algorithms and originators, although we might not have always succeeded in this effort.

A very large group of techniques has evolved in the field of pansharpening, increasing the spatial resolution of a multispectral (MS) image using a panchromatic (PAN) higher-resolution counterpart. Pansharpening, or spatial sharpening belongs to the group of downscaling techniques, like subpixel estimation and regression analysis (Atkinson 2013). Downscaling exists in three aspects, that is, increase in spatial, spectral, and temporal resolution. In image fusion, pansharpening approaches form a subgroup of image fusion techniques available in remote sensing as shown in Figure 4.1.

Image fusion can be used as a tool to increase the spatial resolution. In that case, high-resolution panchromatic imagery is fused with low-resolution often multispectral image data. A distinction has to be made between the pure visual enhancement (superimposition) and real interpolation of data to achieve higher resolution (wavelets), the latter being proposed among others by Ranchin et al. (1996). In this way, the spectral resolution may be preserved while a higher spatial resolution, which represents the information content of the images in much more detail, is incorporated (Franklin and Blodgett 1993; Pellemans et al. 1993). Early examples of fusing SPOT multispectral/panchromatic (XS/PAN) or Landsat TM with SPOT PAN data for resolution enhancement were published, among others, by Cliche et al. (1985), Price (1987), Carper et al. (1990), Franklin and Blodgett (1993), and Ranchin et al. (1996).

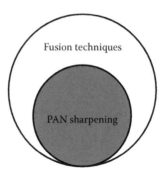

FIGURE 4.1
Pansharpening techniques are a subgroup of remote sensing image fusion algorithms.

A special case forms the fusion of channels from a single sensor for resolution enhancement, for example, TM data. The lower-resolution thermal channel can be enhanced using the higher-resolution spectral channels (Moran 1990). Other approaches increase the spatial resolution of the output channel using a windowing technique on the six multispectral bands of TM (Sharpe et al. 1991). The fusion of SAR/VIR not only results in the combination of disparate data but may also be used to spatially enhance the imagery involved. Geometric accuracy and increase of scales using fusion techniques is of concern to mapping and map updating (Chiesa and Tyler 1990; Pohl 1996).

Pansharpening techniques have now been widely used for some 30 years, since the launch of the SPOT-1 satellite in 1986. With its high-resolution 10-m panchromatic band, remote sensing researchers developed methods of fusing this data with lower-resolution multispectral data, such as the 20-m multispectral bands produced simultaneously by the same satellite and with Landsat multispectral data. The intention was to enhance the spatial resolution of the multispectral imagery, while maintaining the rich spectral information of the multispectral data set (Aiazzi et al. 2012). Pansharpening is valuable to many different applications and has reached a high level of popularity (Aiazzi et al. 2011).

However, to ease understanding, we shall not describe pansharpening techniques in a separate section because many of the developed techniques can be used for the combination of different images for other purposes as well. In this chapter, we have chosen to explain the individual techniques and their consecutive evolution and modifications following a categorization that enables a description of all the various types of image fusion algorithms in remote sensing. This categorization is the result of many years of research and was first published in the *International Journal of Remote Sensing* (Pohl and Genderen 1998) and recently updated in an article published in the *International Journal of Image and Data Fusion* (Pohl and Genderen 2015).

Following the explanation of the categorization, the fusion approaches of each category will be explained in detail. Then a section on how to select input and fusion parameters will follow to allow a quick start into image fusion under selected circumstances. An important section is the summary of research results in terms of the experience with different techniques in order to identify resolved problems on the one hand and contradictions on the other.

4.1 Categorizations of Image Fusion Techniques

It is very important for the understanding of fusion algorithms to have them organized in a system that allows an overview on the existing approaches. The remote sensing community has spent considerable effort to categorize remote sensing image fusion techniques. After having studied all available and implemented pixel-based fusion algorithms in the literature and applied them to two diverse test sites, we published a grouping system that has been widely accepted (Pohl and Genderen 1998). It is still being used in most of the published research and has found its way beyond remote sensing into medical applications as well as military and security domains. Apart from serving as a structure to understand the various algorithms, a good categorization requires that all existing techniques can be accommodated. The grouping system should accommodate all applicable fusion algorithms but contain the least number of necessary categories to avoid complexity (Pohl and Genderen 2015). Sometimes, remote sensing image fusion contains various categories, as for example the principal component substitution (PCS). It has the statistical part to derive principal components from the input image bands, but it also has the component substitution part where the first principal component is replaced by the band to be fused for enhancement, before the data are converted back through an inverse PCA. The decision on where to accommodate a certain technique should be taken based on where the actual fusion process takes place, which would mean for PCS to be accommodated in the group of component substitution techniques.

As a result, we therefore consider six categories of image fusion techniques in the following subsections:

- Component substitution
- Numerical methods
- Statistical image fusion
- Multiresolution approaches
- Hybrid techniques
- Other fusion techniques

The first group comprises algorithms, in which one channel of an image that has been converted into another data space is replaced by an image or band that one wants to fuse. A well-known example is the so-called intensity hue saturation (IHS) fusion explained further below. Numerical methods comprise arithmetic combinations of image bands or different images, including sum, difference, multiplication, and ratio. A popular numerical method that is commonly known in remote sensing is the Brovey transform (BT). Statistical approaches use principal component analysis, a regression analysis, or a least-square fit to merge the data to obtain an optimized fused image. Owing to the need for adaptation, statistical methods are implemented to fit one image channel to the one that is supposed to be replaced or fused. Signal modulation changes one or more characteristics of a wave with a modulating signal that contains relevant information to be introduced. Fusion techniques applying modulation utilize one image or its derivative products to modulate another image for fusion. They are often integrated in other fusion techniques to improve the quality of the outcome. MRAs are very powerful but only have become popular since the increase in computer power that we experienced of the past decades. The fifth and last category of remote sensing image fusion methods, namely, hybrid fusion depicts mixed approaches that incorporate two or more techniques of the four previously mentioned groups. This category obtains increasing interest because it allows a further level of adaptation and specialization to enhance particular information for the desired application of image fusion. It gives more flexibility and takes advantage of the benefits of different techniques in a unified form.

The categorization is not unified in the literature. The groups used in this book have been proven to be useful and are widely accepted. If grouping differs, the authors usually explain the categorization provided in their publication. Sometimes the grouping does not make sense. An example is the relative spatial contribution (RSC) group, which comprises techniques that use a linear combination of spectral bands to fuse the images. This group is categorized as a subgroup of component substitution (CS) pansharpening (Amro et al. 2011). Since there is no substitution, it is more consistent to locate RSC within the numerical techniques category.

4.2 Role of Color in Image Fusion

Before entering the detailed description of the various remote sensing image fusion techniques, it is necessary to explain different color systems and their role in image fusion. It is possible to utilize different color representations to display channels from single or multiple sensor data. Additive primary colors, that is, red, green, and blue allow assigning three different types of information (e.g., image channels). This is called overlay and forms the simplest way of fusion image channels. Owing to the physical principle behind it, it only allows the combination of three image bands. The advantage however

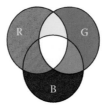

FIGURE 4.2
Overlay model of primary colors red, green, and blue in relation to cyan, magenta, and yellow.

lies in its simplicity and effectiveness, especially in applications like geology or flood damage assessment where the combination of complementary optical and radar images plays an important role.

Additive *primary colors* allow the assignment of three different types of information (e.g., image channels) to the three primary colors red, green, and blue. Together they form a color composite that can be displayed with conventional media, for example, cathode ray tube. Figure 4.2 shows the primary red, green, and blue color system and their relationship to cyan, magenta, and yellow, which are mainly used in printing.

Another representation is the color cube as depicted in Figure 4.3. The brightness levels of red, green, and blue create the color cube. For an image representation, each pixel consists of three values, which form the coordinates in this color cube. Each of the three values is retrieved from the image bands assigned to the red, green, and blue channel. The points on the *gray line* represent the equal contribution from the three primary colors. The color cube also allows the representation of the three colors used in printing, that is, cyan, magenta, and yellow.

A color composite facilitates the interpretation of multichannel image data due to the variations in colors based on the values in the single channels.

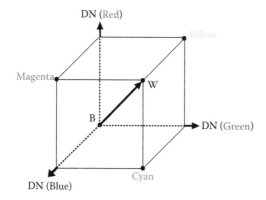

FIGURE 4.3
Color cube of the RGB primary color system.

The gray scale value, used to select a set of red, green, and blue brightness, is stored in a look-up table (LUT) that are the voltages sent to the display tube. Operations on the LUT and the histogram of the image data can enhance the color composite for visual interpretation.

The possibilities of varying the composite are manifold. Depending on the selection of the input image channels, the fused data will show different features. Very important for the color composite is the distribution of the available 256 (0–255) gray values to the range of the data in the case of an 8-bit radiometric resolution. It might be of advantage to invert input channels before combining them in the RGB display with other data depending on the objects of interest to be highlighted. In many cases, the RGB technique is applied in combination with another image fusion procedure, for example, IHS, PCA, and others, which are explained in the sections below.

For the fusion process using different color spaces, the graphical representation of the relationship between RGB and IHS might be helpful for further understanding the processes required. The transformation from RGB to IHS can be depicted as *cylinder* (see Figure 4.6) or in a *hexcone model* (Figure 4.4). For the latter, the RGB color cube is projected onto a plane positioned perpendicular to the gray one. The result is a hexagon. The distance along the line in the direction of the white point defines the intensity (I). Hue (H) is obtained by the angle around the hexagon, while the saturation (S) of the color is the distance from the gray point at the center of the hexagon. Using the points A, B, G, P, and P' as shown in Figure 4.4, the values for H and S can be calculated from Equations 4.1 and 4.2, respectively. The hexcone model illustrates the RGB to IHS mathematical relationship (Schowengerdt 2007a).

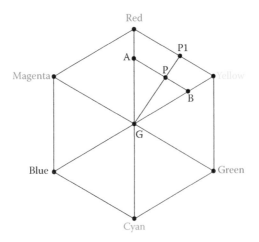

FIGURE 4.4
Graphical representation of the RGB to IHS transform as hexagon.

$$H = \frac{\overline{AP}}{\overline{AB}} \qquad (4.1)$$

$$S = \frac{\overline{GP}}{\overline{GP'}} \qquad (4.2)$$

Especially in the case of fusing images to increase the spatial resolution of multispectral data, it might be necessary to match the spectral characteristics of the high-resolution panchromatic to each multispectral image band. Often, the panchromatic channel has to be matched to a transformed image channel, such as the intensity or the first principal component. If the matching is done, spectral distortions can be minimized in the fused image. The matching is performed on the basis of the mean and standard deviation of the bands to be matched as indicated in Equation 4.3, adapted from Amro et al. (2011):

$$\text{PAN}_j^* = (\text{PAN}_j - \mu_{\text{PAN}}) \frac{\sigma_{\text{MS}i}}{\sigma_{\text{PAN}}} + \mu_{\text{MS}i} \qquad (4.3)$$

in which PAN_j^* is the histogram-matched pixel of the panchromatic image pixel j, and μ_{PAN} and $\mu_{\text{MS}i}$ are the means of the panchromatic image and band i of the multispectral image MS, respectively. σ_{PAN} and $\sigma_{\text{MS}i}$ are the standard deviations of PAN and MS in band i of the multispectral image, respectively.

4.3 Component Substitutions

CS is very popular due to its simplicity. Another term for this group that can be found in the literature is projection substitution (PS) (Thomas et al. 2008). In fact, it comprises elements of other groups of fusion techniques further explained below. Strictly speaking, the category of component substitution techniques is a "super-group," containing color transform, statistical approaches, as well as multiscale decompositions described in Sections 4.3.1, 4.5.3, and 4.6. That is one of the reasons why published remote sensing image fusion research is difficult to evaluate objectively; as each group of authors tends to categorize the methods they analyzed based on subjective criteria to suit their personal selection.

The idea of CS techniques is to convert a number of bands of the original image into another data space (e.g., another color space) where one of the resulting channels is replaced by a new image (e.g., higher spatial resolution image). The reverse transform creates the actual fused image, containing information from both input data. Therefore, these techniques are also called projection techniques (Thomas et al. 2008). The described general approach of CS is depicted in Figure 4.5. It is possible to implement CS without explicitly running the transformations using a projection scheme (Tu et al. 2001). The most popular CS approach is the IHS transform, which is the first method that arises when remote sensing image fusion is mentioned. Sometimes, researchers call it HSI (hue saturation intensity) or HSV (hue saturation value) or use another color description as in the hyperspherical color space (HCS) (Padwick et al. 2010). The HCS is a model that is found to facilitate color discrimination following the preferences of human vision. The difference between two colors is the Euclidean distance between the colors that are distributed on the surface of a hypersphere in a four-dimensional space (Leonov and Sokolov 2008).

IHS comes with a series of adaptations resulting in new names. IHS and its derivatives are explained in detail in the next subsection. Another very popular CS technique is using a PCA. When used as CS, this technique is often called PCS. Strictly speaking, PCS is a hybrid approach because it uses a statistical approach and CS.

A CS approach invented by Laben and Bower and later patented by Eastman Kodak uses simulated panchromatic data (Laben et al. 2000). It is called Gram–Schmidt (GS) fusion and classified as a CS method. The multispectral and a simulated panchromatic band are transformed by the GS transform (GST), and the resulting bands with the original matched panchromatic image are used for the CS approach. It will be explained further in Section 4.3.3, where a graphical representation is provided as well (Figure 4.11).

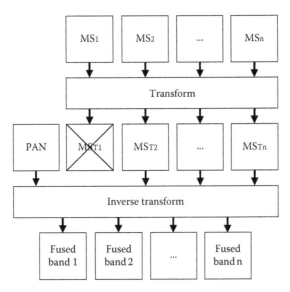

FIGURE 4.5
Generic workflow for a CS image fusion.

4.3.1 Intensity Hue Saturation Transform

Apart from Brovey (see Section 4.4.4), IHS is one of the most popular techniques in remote sensing image fusion. IHS uses a mathematical color model as explained above using a cylindrical or spherical coordinate system. The equations vary slightly depending on the underlying color system. Al-Wassai et al. (2011) provide a very detailed overview on the various models and their impact on the fusion process for further reading. The IHS color transformation effectively separates spatial (I) and spectral (H, S) information from a standard RGB image. It relates to the human color perception parameters. A very commonly used mathematical context is expressed by Equations 4.4 through 4.7, based on a cylindrical color model (see Figure 4.6). I relates to the intensity, while v_1 and v_2 represent intermediate variables, which are needed in the transformation. H and S stand for hue and saturation, respectively.

$$\begin{pmatrix} I \\ v_1 \\ v_2 \end{pmatrix} = \begin{pmatrix} \dfrac{1}{\sqrt{3}} & \dfrac{1}{\sqrt{3}} & \dfrac{1}{\sqrt{3}} \\ \dfrac{1}{\sqrt{6}} & \dfrac{1}{\sqrt{6}} & -\dfrac{2}{\sqrt{6}} \\ \dfrac{1}{\sqrt{2}} & -\dfrac{1}{\sqrt{2}} & 0 \end{pmatrix} * \begin{pmatrix} R \\ G \\ B \end{pmatrix} \tag{4.4}$$

$$I = \frac{(R+G+B)}{3} \tag{4.5}$$

$$H = \tan^{-1}\left(\frac{v_2}{v_1}\right) \tag{4.6}$$

$$S = \sqrt{v_1{}^2 + v_2{}^2} \tag{4.7}$$

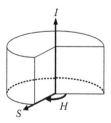

FIGURE 4.6
IHS cylindrical color model.

The standard flow of the IHS transform for image fusion is shown in Figure 4.7. There are two ways of applying the IHS technique in image fusion: direct and substitutional. The first refers to the transformation of three image channels assigned to I, H, and S. The second transforms three channels of the data set representing RGB into the IHS color space, which separates the color aspects in its average brightness (intensity). This corresponds to the surface roughness, its dominant wavelength contribution (hue) and its purity (saturation). Both the hue and the saturation in this case are related to the surface reflectivity or composition. Then, a fourth image channel replaces one of the components.

In many published studies, the channel that replaces one of the IHS components is contrast stretched to match the latter. A reverse transformation from IHS to RGB as presented in Equation 4.8 converts the data into their original image space to obtain the fused image:

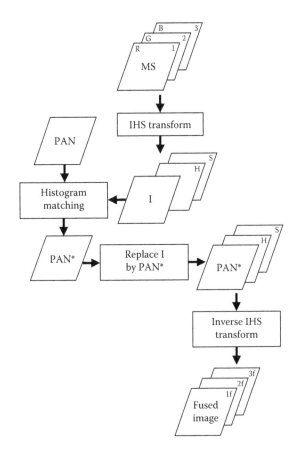

FIGURE 4.7
Workflow for IHS fusion.

$$
\begin{pmatrix} R \\ G \\ B \end{pmatrix} = \begin{pmatrix} \dfrac{1}{\sqrt{3}} & \dfrac{1}{\sqrt{6}} & \dfrac{1}{\sqrt{2}} \\ \dfrac{1}{\sqrt{3}} & \dfrac{1}{\sqrt{6}} & -\dfrac{1}{\sqrt{2}} \\ \dfrac{1}{\sqrt{3}} & -\dfrac{2}{\sqrt{6}} & 0 \end{pmatrix} * \begin{pmatrix} I \\ v_1 \\ v_2 \end{pmatrix}
\tag{4.8}
$$

The IHS technique has become a standard procedure in image analysis. It serves color enhancement of highly correlated data, feature enhancement, the improvement of spatial resolution, and the fusion of disparate data sets. The use of the IHS technique in image fusion is manifold, but based on one principle: the replacement of one of the three components (I, H, or S) of one data set with another image. Most commonly, the intensity channel is substituted. Replacing the intensity—the sum of the bands divided by the number of bands as given in Equation 4.5—by a higher spatial resolution value and reversing the IHS transformation leads to composite bands (Equation 4.8). These are linear combinations of the original (resampled) multispectral bands and the higher-resolution panchromatic band.

A variation of the IHS fusion method applies a stretch to the hue and saturation components before they are combined and transformed back to RGB. This is called color contrast stretching (Gillespie et al. 1986). The IHS transformation can be performed either in one or in two steps. The two-step approach includes the possibility of contrast stretching the individual I, H, and S channels. It has the advantage of resulting in color enhanced fused imagery.

The limitation of IHS to three bands was resolved by Tu et al. (2001) who introduced the generalized IHS (GIHS). They used a nonlinear RGB-IHS transformation expressed in Equations 4.5 and 4.9 through 4.11:

$$
H = \begin{cases} \cos^{-1}(a) & \text{if } G \geq R \\ 2\pi - \cos^{-1}(a) & \text{if } G \leq R \end{cases}
\tag{4.9}
$$

$$
a = \frac{(2B - G - R)/2}{\sqrt{(B-G)^2 + (B-R)(G-R)}}
\tag{4.10}
$$

$$
S = 1 - \frac{3\min(R, G, B)}{R + G + B}
\tag{4.11}
$$

The fusion process can then be rewritten as Equation 4.12:

$$
\begin{pmatrix} R_{\text{fused}} \\ G_{\text{fused}} \\ B_{\text{fused}} \end{pmatrix} = \begin{pmatrix} R + \delta \\ G + \delta \\ B + \delta \end{pmatrix}
\tag{4.12}
$$

with

$$\delta = PAN - I \tag{4.13}$$

Therefore, any number of bands can be fused if Equations 4.12 and 4.13 are generalized to Equation 4.14:

$$MS_{\text{fused } i} = MS_i + \beta \tag{4.14}$$

where β equals the difference between the high-resolution PAN image and each individual multispectral band i of n number of bands. It is calculated using Equation 4.15:

$$\beta = \frac{PAN - 1}{n \sum_{i=1}^{n} MS_i} \tag{4.15}$$

The reason why so many adapted versions of IHS have been developed since its first occurrence is the problem of spectral discrepancy between the panchromatic band and the multispectral image channels. This causes a deterioration of the spectral content in the fused image. In particular, the saturation value is influenced by IHS-based methods. There are various solutions to this problem, which we explain in the following paragraphs.

The modified IHS (modIHS) introduces the near-infrared band as an auxiliary input to filter out the effect of near-infrared reflectivity on the panchromatic image (Siddiqui 2003). The trade-off between spatial improvement and spectral quality loss has received much attention and led to the introduction of trade-off parameters (Choi 2006; Te-Ming et al. 2007). These parameters allow a fine tuning by the user in order to obtain the desired result.

Continuing the advancement of IHS-based methods to overcome spectral quality problems, researchers have proposed the image- and edge-adaptive IHS (AIHS) (Rahmani et al. 2010), the spectrally adjusted IHS (SAIHS) (Te-Ming et al. 2004), and the improved AIHS (IAIHS) (Leung et al. 2014). The latter modification appeared necessary to still overcome spectral distortions. They proposed to combine MS-induced with PAN-induced weights to improve the spectral quality of the fused image while maintaining spatial detail. Again, this is an approach following an adaptive trade-off between the two qualities.

An algorithm combining IHS with an FFT where an enhancement of the spatial information is performed in the frequency domain provides a solution to the spectral quality deterioration. This led to a hybrid method called Ehlers fusion, which will be explained in Section 4.7.1.

4.3.2 Principal Component Substitution

PCS is based on PCA, also known as the *Karhunen–Loève* transform, and converts input channels into a set of uncorrelated orthogonal components. The transformation uses the eigenvectors of the covariance matrix of the input image. The actual fusion process follows the component substitution scheme, replacing the first principal component (PC_1) by the histogram-matched PAN image. A reverse PCA returns the high-resolution fused image in original coordinate system or color space. This works because PC_1 contains the information that is common to all bands, that is, the spatial information content of the image channels, while the spectral information remains in the other principal components. The procedure is displayed in Figure 4.8.

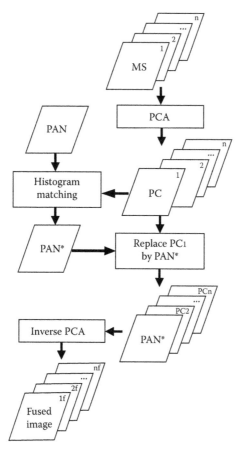

FIGURE 4.8
PCS image fusion processing steps.

PCA is useful for image encoding, image data compression, image enhancement, digital change detection, multi-temporal dimensionality, and image fusion. It can be categorized as a statistical technique because it transforms a multivariate data set of intercorrelated variables into a data set of new uncorrelated linear combinations of the original variables. It generates a new set of axes, which are orthogonal.

PCA is best understood when looking at a two-band scatterplot, which presents the distribution of pixel values of the two bands in a two-dimensional space (compare with Figure 4.9). Each plotted point has a coordinate: the pixel value of band A in the x-direction and the value of band B in the y-direction. An example is presented in Figure 4.9 where the plotted point P has the coordinates P(98,65), meaning that the values in band A or $x = 98$ and band B or $y = 65$. If data of the two bands follow a normal distribution, the resulting shape is an ellipse. Adding more than three bands, the dimension of the ellipse turns into a hyper-ellipsoid.

The idea of principal components is now to establish new spectral data axes in parallel to the main axes of the ellipse/ellipsoid. Like this, the resulting first principal component (PC_1) will correspond to the major axis of the ellipse, which is the longest of all, also shown in Figure 4.10. Two other parameters need to be defined, that is, the eigenvector and the eigenvalue. The eigenvector of PC_1 is the direction of PC_1: the length represents the first eigenvalue. With the new coordinate system, the pixels of the different bands obtain new values that follow the new axes. Owing to its nature, PC_1 represents the highest variation in the data. All subsequent components are perpendicular to the first. Considering n bands of the input data, it will result in n PCs. Figure 4.3 displays PC_1 and PC_2 for a two-dimensional (2D) case.

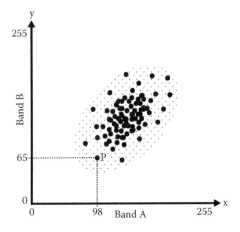

FIGURE 4.9
Scatterplot of a two-band image (2D).

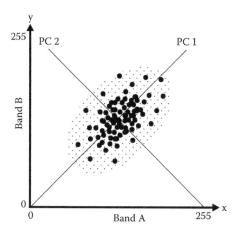

FIGURE 4.10
Location of the axes of the new coordinate system representing PC_1 and PC_2 in the 2D case.

The approach for the computation of the principal components (PCs) comprises the calculation of the

1. Covariance (nonstandardized PCA) or correlation (standardized PCA) matrix
2. Eigenvalues—vectors
3. PCs

The covariance matrix contains all the variances and covariances within the different image bands. The size of the matrix is $n \times n$, with n being the number of bands involved. It has to be noted that the covariances between similar bands are the same. The diagonal contains the band variances because the covariance of a band with itself is equal to its variance. Table 4.1 shows an example of a 3×3 covariance matrix of a three-band image for illustration.

For the determination of the variance σ_A^2, the mean μ_A of the gray values of a particular image band A is needed. It is determined as a statistical average of k number of pixels by Equation 4.16:

TABLE 4.1

Covariance Matrix Example for an Image with Three Bands

	Band 1	Band 2	Band 3
Band 1	V_1	CoV_{21}	CoV_{31}
Band 2	CoV_{12}	V_2	CoV_{32}
Band 3	CoV_{13}	CoV_{23}	V_3

$$\mu_A = \frac{\sum_{j=1}^{k} A_j}{k} \tag{4.16}$$

The variance describes how much the different pixel values vary in one band. It is calculated by Equation 4.17, in which j stands for a particular pixel; A_j is the gray value of that pixel.

$$\sigma_A^2 = \frac{\sum_{j=1}^{k} (A_j - \mu_A)^2}{k-1} \tag{4.17}$$

The covariance describes the relationship between two image bands. It represents the trend of gray values in a certain pixel across different bands in relation to the means of their respective bands. The covariance is the average product of the differences of corresponding values in two different bands from their respective means. Equation 4.18 calculates the sample covariance C_{AB} with A and B representing the gray values of two different image bands at a pixel j of k number of pixels and μ_A, μ_B the band means for band A and B, respectively:

$$C_{AB} = \frac{\sum_{j=1}^{k} (A_j - \mu_A)(B_j - \mu_B)}{k} \tag{4.18}$$

PCA transforms the data so that the axes in the n-dimensional spectral space are translated and rotated to match the new axes of the ellipse representing the principal components. This is done using a linear transformation as shown in Equation 4.19, which needs eigenvectors and eigenvalues that are calculated from the covariance matrix with Equation 4.20:

$$V = E \operatorname{Cov} E^T \tag{4.19}$$

In this equation, Cov is the covariance matrix and E the matrix of eigenvectors that also appears in the transposed form E^T. The resulting matrix of eigenvalues V contains only values along the diagonal. All nondiagonal elements are zeros. In order to obtain the greatest variance in the first PC, it is necessary that the eigenvalues, which represent the variance values for each PC along the diagonal in V, are organized from the largest to the smallest value $(v_1 > v_2 > v_3 ... > v_n)$. The columns in the eigenvector matrix E contains the coefficients to transform the original image values into the PC values with Equation 4.20 (Smith et al. 1999):

$$P_e = \sum_{i=1}^{n} p_i E_{ie} \qquad (4.20)$$

with e being the number of the PC, P_e the output value in the principal component calculated from band i, and n the number of input bands containing pixel value p.

An inverse PCA transforms the combined data back to the original coordinate system. The use of the correlation matrix implies a scaling of the axes so that the features receive a unit variance. It prevents certain features from dominating the image because of their large digital numbers. The signal-to-noise ratio (SNR) can significantly be improved applying the standardized PCA. Better results are obtained if the statistics are derived from the whole study area rather than from a subset area.

PCA in image fusion has two approaches:

1. PCA of multichannel image—replacement of first principal component by different images (*principal component substitution*—PCS) (see Section 4.5.3)
2. PCA of all multi-image data channels

The first version follows the idea of increasing the spatial resolution of a multichannel image by introducing an image with a higher resolution. The channel, which will replace PC_1, is stretched to the variance and average of PC_1. The higher-resolution image replaces PC_1 since it contains the information, which is common to all bands while the spectral information is unique for each band; PC_1 accounts for maximum variance, which can maximize the effect of the high-resolution data in the fused image.

The second procedure integrates the disparate natures of multi-sensor input data in one image and applies a PCA. It will be further discussed in Section 4.5.3.

The PCA approach is sensitive to the choice of area to be analyzed. The correlation coefficients reflect the tightness of a relationship for a homogeneous sample. However, shifts in the band values due to markedly different cover types also influence the correlations and particularly the variances.

4.3.3 Gram–Schmidt Transform

The original GS orthogonalization process, named after two mathematicians, even though they did not exactly develop it, is a method that orthonormalizes a set of vectors in a given space. It was patented as GS spectral sharpening (Laben et al. 2000). As a generalization of PCS, it could be considered a statistical method since it de-correlates the input bands using their covariance values. In image fusion, it is applied to transform the resampled MS image and a simulated panchromatic band by GST as each image channel corresponds to one high-dimensional vector. The histogram-matched panchromatic band replaces the first GS component and a reverse transform produces the actual fused image. Figure 4.11 illustrates the processing flow of the Gram–Schmidt image fusion.

First, the PAN is simulated using a linear combination of the n bands of the MS image described in Equation 4.21:

$$PAN' = \sum_{i=1}^{n} w_i \, MS_i \qquad (4.21)$$

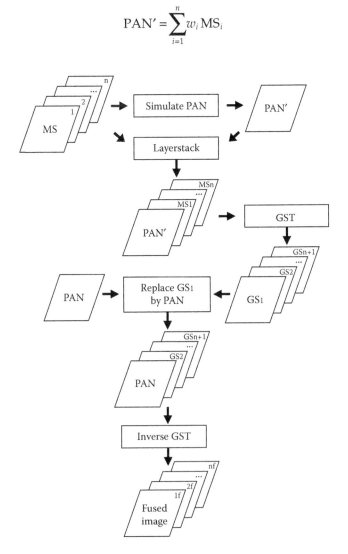

FIGURE 4.11
GS image fusion.

with PAN' being the simulated panchromatic image, $i = \{1, 2, \ldots, n\}$ at n number of multispectral MS bands. Each band is considered a high-dimensional vector choosing PAN' as the first vector. Then GST will produce a set of vectors that are orthogonal to each other and normalized. It applies the projection operator Proj_a for the vectors a and b and their inner product $a|b$ as depicted in Equation 4.22:

$$\text{Proj}_a(b) = \frac{\langle a|b \rangle}{\langle a|a \rangle} a \tag{4.22}$$

This operator projects the vector b orthogonally onto the line spanned by vector a. The Gram–Schmidt process follows the steps noted in Equations 4.23 through 4.26 below:

$$a_1 = b_1 \tag{4.23}$$

$$a_2 = b_2 - \text{Proj}_{a1}(b_2) \tag{4.24}$$

$$a_3 = b_3 - \text{Proj}_{a1}(b_3) - \text{Proj}_{a2}(b_3) \tag{4.25}$$

$$a_m = b_m - \sum_{j=1}^{m-1} \text{Proj}_{a_j}(v_m) \tag{4.26}$$

where $a_1, \ldots, a_m$ produces the orthogonal vectors, which are normalized then into ortho-normalized $c_1, \ldots, c_m$ vectors using Equations 4.27 through 4.30 with m as the number of MS bands plus the panchromatic channel, that is, $m = n + 1$.

$$c_1 = \frac{a_1}{\|a_1\|} \tag{4.27}$$

$$c_2 = \frac{a_2}{\|a_2\|} \tag{4.28}$$

$$c_3 = \frac{a_3}{\|a_3\|} \tag{4.29}$$

$$c_m = \frac{a_m}{\|a_m\|} \tag{4.30}$$

In the fusion process, it is an iterative procedure that computes the angle between the input band and the panchromatic image, and rotates the band to make it orthogonal to the PAN.

The actual fusion takes place after the forward GST. The low-resolution simulated PAN' is replaced by the gain and bias adjusted high-resolution PAN band. All multispectral bands are resampled to the same resolution. Using the coefficients on the high-resolution channels, the reverse GST converts the fused image back into its original space.

Variations of GS exist based on the method to produce the synthetic PAN at low resolution. GS uses averaging of the MS components (mode 1) or a low-pass filter (mode 2) (Laben et al. 2000). Adaptive GS (GSA) uses a weighted average using mean square error (MSE) minimizing weights, resulting in the algorithm called minimum MSE (MMSE) (Garzelli et al. 2008). Other authors consider it a drawback that there is so much variety of methods that are used for the calculation of weights to simulate the low-resolution panchromatic band from the multispectral data. They can be estimated from the sensor's spectral sensitivity curves, using linear regression or least-square methods. A straightforward solution to this problem is the use of the covariance matrix of the MS to calculate the PAN weights (Maurer 2013).

4.4 Numerical Methods

Mathematical combinations of different images are among the simplest and earliest methods used in remote sensing. Multiplicative approaches, difference images, and ratios play an important role in Earth observation. An example is the subtractive resolution merge (SRM), just one of the many others we describe in the coming sections.

Numerical methods are the simplest approaches that we can follow since they combine the digital numbers of the images to be fused at a pixel-by-pixel basis using mathematical operators such as sum, multiplication, subtraction (or differencing), and division (also called ratio). Any combination of these operators is possible as shown in the description of the commonly used algorithms below.

4.4.1 Sum

The addition of multispectral and panchromatic image channels found its implementation including a high-pass filter (HPF) to increase the spatial feature of the high-resolution panchromatic image in the fused result. As HPF fusion method, this approach can be found in the literature (Schowengerdt 1980; Chavez and Bowell 1988; Chavez et al. 1991). Using the ratio of the spatial resolution of the two input images, the size of the kernel is defined for the high-pass convolution filter applied to the panchromatic image. The resulting sharpened image is added to the multispectral bands, including a weighting factor depending on the standard deviation of the multispectral bands toward the panchromatic channel. A linear stretch adapts the fused image to the original multispectral data. Further information about HPF fusion can be found in Section 4.4.7.

4.4.2 Multiplication

Multiplication can be a very powerful fusion method and leads to very good results for visual interpretation if optical and radar images are combined (Pohl and Genderen 1995). Fusion by multiplication enhances the contrast and joins multispectral with textural information from the input images. An example of a multiplication process is expressed in Equation 4.31 (Yésou et al. 1993b):

$$DN_f = A(w_1 DN_a + w_2 DN_b) + B \tag{4.31}$$

A and B are scaling factors and w_1 and w_2 weighting parameters. DN_f, DN_a, and DN_b refer to digital numbers of the final fused image and the input images a and b, respectively. The choice of weighing and scaling factors may improve the resulting images. Some authors refer to the multiplication method as intensity modulation (IM) in the context of pansharpening. PAN is used to intensity modulate each multispectral band separately. The mathematical expressions for IM are given in Equations 4.32 through 4.34:

$$XS_1^H = a_1 + b_1\sqrt{P * XS_1^L} \tag{4.32}$$

$$XS_2^H = a_2 + b_2\sqrt{P * XS_2^L} \tag{4.33}$$

$$XS_3^H = a_3 + b_3\frac{P + 3XS_3^L}{4} \tag{4.34}$$

in which XS_i^H refers to the fused high-resolution multispectral channels, XS_i^L is the original multispectral band ($i = \{1,2,3\}$), and P is the high-resolution panchromatic band (Bretschneider and Kao 2000). The square root of the mixed data accounts for the increased brightness values and reduces the data to the original gray value scale.

4.4.3 Subtractive Resolution Merge and Ratioing

This method, commonly called image differencing, uses two images acquired at two different points in time. The images are subtracted from each other on a pixel by pixel basis. In change detection, it is known as univariate image differencing (Singh 1989). In this context, the subtraction can take place after additional transformations of the original image into indices. A variation of this approach is the change vector analysis (CVA), which allows the simultaneous processing of all channels of the input images. Spectral vectors are subtracted pixel by pixel. They can be preprocessed prior to fusion to enhance desired features. CVA delivers the change magnitude and change direction (Coppin et al. 2004). The difficulty is to decide where to put the threshold between "change" and "no change." Difference or ratio images are very suitable for change detection (Singh 1989).

The ratio method, where the images are ratioed instead of subtracted, is even more useful because of its capability to emphasize more on the slight signature variations (Zobrist et al. 1979; Singh 1989). The advantage is that illumination differences that are caused by topography can be eliminated. In some cases, the resulting difference image contains negative values. Therefore, a constant C has to be added to produce positive digital numbers. However, differences do not always refer to changes since other factors, like differences in illumination, atmospheric conditions, sensor calibration, ground moisture conditions, and registration of the two images, can lead to differences in radiance values. In ratioing, two images from different dates are divided, band by band if the image data have more than one channel. If the intensity of the reflected energy is nearly the same in each image, then the ratio image pixel is one; it indicates no change. The critical part of this method is selecting appropriate threshold values in the lower and upper tails of the distribution representing change pixel values. In this respect, the normalization of the data is of advantage as indicated in Equation 4.35:

$$DN_f = \frac{XS_3 - XS_2}{XS_3 + XS_2} - \frac{TM_4 - TM_3}{TM_4 + TM_3} + C \tag{4.35}$$

A ratio for spatial enhancement is summarized by Munechika et al. (1993). The aim of this method is to maintain the radiometric integrity of the data while increasing the spatial resolution. Equation 4.36 is given here:

$$DN_f = DN_P - \frac{DN_{XSi}}{DN_{synP}} \tag{4.36}$$

where DN_f represents the fused high-resolution image, DN_P stands for the corresponding pixel in the high-resolution image, DN_{XSi} the input low-resolution multispectral image (*i*th band), and DN_{synP} is the corresponding pixel in the low-resolution synthesized panchromatic image, created from the low-resolution multispectral bands that overlap the spectral response of the input high-resolution panchromatic band.

That is also the aim of the Brovey transform, as discussed below, a formula that normalizes multispectral bands used for an RGB display, and multiplies the result by any other desired data to add the intensity or brightness component to the image.

Popular further processing of the individual images prior to ratioing is the production of various spectral index ratios (SIR), such as landmass (NDLI), vegetation (NDVI), water (NDWI), or snow (NDSI) indices. ND stands for normalized difference, and I for index. The SIR is used to classify a particular object. It is directly related to the difference in reflectance values of the bands that form the ratio (Jawak and Luis 2013). If SIRs are used for subtraction, the approach is called index differencing.

4.4.4 Brovey Transform

One arithmetic combination has managed to survive until today: the BT. It is a classical technique and one of the most successful methods, based on spectral modeling that reaches a normalization of the input bands through addition, subtraction, and ratio. Its follower color normalized spectral sharpening (CN), described in the next section, enables the incorporation of any number of bands (Hallada and Cox 1983) rather than the limiting three originally introduced by Brovey. Named after its author, the Brovey transform normalizes multispectral bands and multiplies the resulting channels with the intensity or brightness channel, which could be a panchromatic image. It is not exactly a transform but a multiplication using a panchromatic band based on a normalization of the multispectral bands. That is why it is mentioned in this category of arithmetic combinations. BT preserves the relative spectral contribution of each pixel. It replaces the overall brightness with the image channel to be fused (e.g., a high-resolution panchromatic or a SAR image). It was originally developed to visually increase contrast, especially at both ends of the histograms. Therefore, the often-mentioned disadvantage of BT not preserving the spectral quality of the original image is not justifiable because it was not designed to do so. It still remains a very important technique that has been amended to accommodate more than three channels, and also found its way into hybrid approaches discussed later on.

The algorithm behind it is depicted in Equation 4.37 with BT_i stands for the resulting digital number digital number of the fused image band i with n number of multispectral bands. PAN is the digital number of the image to be fused to either improve spatial resolution (e.g., by a high-resolution panchromatic image) or to combine optical with radar data using a SAR image (adapted from Pohl and Genderen 1998):

$$BT_i = \frac{MS_i}{\sum_i^n MS_i} PAN \tag{4.37}$$

For $i = \{1, 2, 3\}$, we obtain the equation representing the Brovey trans-form. However, by using weights and a near-infrared (NIR) band, BT can be adapted to suit certain application that need the integration of a NIR chan-nel. The advanced algorithm is depicted in Equation 4.38 where w_x is the weighting factor with $x = \{1, 2, 3, \text{NIR}\}$ and $i = \{1, 2, 3, \text{NIR}\}$:

$$BT_{i\,\text{NIR}} = MS_i \frac{\text{PAN} - w_{\text{NIR}} * \text{NIR}}{w_{b1}MS_1 + w_{b2}MS_2 + w_{b3}MS_3} \qquad (4.38)$$

with $BT_{i\,\text{NIR}}$ representing the resulting fused image channel.

4.4.5 Modulation-Based Techniques

First introduced by Liu (2000) as smoothing filter-based intensity modula-tion (SFIM), this approach uses a ratio between the PAN and its low-pass fil-tered image to modulate a lower resolution MS image. The approach follows the following Equation 4.39:

$$\text{SFIM}_{i,j,k} = \frac{MS_{i,j,k} * \text{PAN}_{i,j}}{\text{Mean}_{i,j}} \qquad (4.39)$$

where $\text{SFIM}_{i,j,k}$ represents the fused output image, and i, j represent the pixel location in band k. Mean stands for the simulated low-resolution panchro-matic image, where neighboring pixels are averaged using a smoothing fil-ter. The window size for the filter is derived from the spatial ratio of MS and PAN.

Khan et al. (2008) improved the algorithm by upscaling the MS image using a nonlinear interpolation instead of bicubic interpolation. Their tech-nique is called *Indusion* in the literature. To make it applicable to VIR/SAR fusion, Alparone et al. (2004) use a modulation of the MS-derived intensity channel with spatial detail.

The interband structure model (IBSM) is designed to adjust the resulting fused image closely to what an MS image would look like if it had the high spatial resolution of the PAN. It uses information of gain and offset of spatial structures when injecting this information from PAN to MS (Garzelli and Nencini 2005) and should consider the modulation transfer function (MTF). Aiazzi et al. (2003) mentioned that any technique using spatial information injection should consider the MTF of the imaging system to avoid spectral distortion. IBSM finds its application in WT-based fusion approaches.

4.4.6 Color Normalized Spectral Sharpening

CN is a further development of the Brovey transform to allow more than three multispectral input channels. The algorithm normalizes the data and separates the spectral space into hue and brightness. It multiplies each multispectral channel by the high-resolution panchromatic image and divides the result by the sum of the input multispectral bands, which causes the normalization. The equation for CN is represented in Equation 4.40, with CN_i as the fused color normalized sharpened spectral band i, and n the number of spectral input bands. The constants 1 and 3 have to be added to avoid a division by zero.

$$CN_i = \frac{3(MS_i + 1)(PAN + 1)}{\sum_i^n MS_i + 3} - 1 \tag{4.40}$$

4.4.7 High-Pass Filtering

The HPF resolution merge is one of the first established image fusion methods used (Schowengerdt 1980, 2007b,c; Chavez et al. 1991). It has survived 30 years of research in image fusion. The idea of injecting spatial detail extracted from the higher-resolution PAN or SAR into lower-resolution MS images is based on this method and has further developed over the years. In the literature, it can be found as HPF, HPF addition (HPFA) (Gangkofner et al. 2008) or high-frequency injection (HFI) (Schowengerdt 1980). The injection plays an important role in all MRA approaches mentioned in Section 4.6, which is why in a generic perspective, HPF could be categorized under MRA (Vivone et al. 2015). It also finds its application in hybrid approaches such as the Ehlers fusion.

The method follows three steps (see Figure 4.12):

1. High-pass filtering the high-resolution PAN image.
2. Add the high-pass filtered image to each multispectral band using individual weights depending on the standard deviation of the MS bands.
3. Match the histograms of the fused image to the original MS bands.

The principle of HPF is to enhance the spatial content of the high-resolution panchromatic image using a high-pass filter. This extracts the high-frequency information, which is then added to each individual multispectral band of the MS image. For a successful implementation of the HPF algorithm, the size of the filter kernel has to be determined. It depends on the ratio R of the pixel resolutions PR of the two input images as depicted in Equation 4.41:

$$R = \frac{PR_{MS}}{PR_{PAN}} \tag{4.41}$$

with PR_{MS} corresponding to the multispectral and PR_{PAN} to the panchromatic image. The identified optimal kernel size is $2R$.

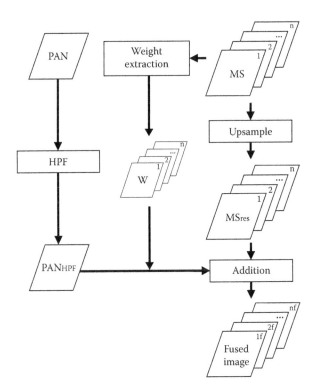

FIGURE 4.12
Processing steps for HPF fusion.

An improved result is obtained if the HPF enhanced panchromatic image PAN_HPF is weighted before added to all the MS bands. The individual weights w_i are obtained from the global standard deviation of the related MS$_i$ band. The standard deviation σ_{MSi} for band MS$_i$ is retrieved using the square root of the band variance in Equation 4.42:

$$\sigma_{MSi} = \sqrt{\frac{\sum_{j=1}^{k}(MS_{ij} - \mu_{MSi})^2}{k-1}} \tag{4.42}$$

with j referring to a particular pixel; MS$_{ij}$ is the gray value of that pixel in band i of the multispectral image. The equation needs the mean μ_{MSi} of the gray values of the particular image band i. The weight w_i is obtained from Equation 4.43:

$$w_i = \frac{\sigma_{MSi}}{\sigma_{PAN\_HPF}} f \tag{4.43}$$

whereby f stands for a modulation factor that allows user intervention, leading to a crisper result.

The fusion process uses Equation 4.44:

$$\text{MS}_{\text{HPF}} = \text{MS}_{res} + \text{PAN}\_\text{HPF}w \tag{4.44}$$

with MS_{HPF} as the resulting HPF fused multispectral image, MS_{res} the resampled original multispectral image to match PR_{PAN}, PAN_HPF the high-pass filtered panchromatic image, and w the calculated weights.

A modification of HPF multiplies the PAN image with each band of the MS imagery, followed by a normalization using the low-pass filtered PAN to create the fused image. This method is called high-pass modulation (HPM) or high-frequency modulation (HFM) (Schowengerdt 2007b).

4.5 Statistical Image Fusion

This group of method uses statistical relationships between the input images in the fusion process, for example, depending on the correlation between image pixels of both images. If this relationship changes depending on the location, the algorithm is called adaptive. In the process of achieving better results for the fused image, researchers moved on to calculating a linear regression to extract the spectral contribution of both, MS and PAN. However, as pointed out by Thomas et al. (2008), the relationship between both is not at all linear. To solve this problem, the histograms of the input images are adjusted so that the means and standard deviations have the same values (Xu et al. 2015) and the process has been converted into a spatially adaptive approach.

4.5.1 Spatially Adaptive Image Fusion

One of the first statistical image fusion methods proposes an adaptive insertion of information depending on the local correlation of the two input images, for example, PAN and MS (Price 1999). It was further developed to avoid block artifacts especially in the areas of different spectral characteristics. The nonlinear relationship of high-frequency information in the two input images had to be modeled. The improvement was achieved by introducing a spatially adaptive window to estimate the gain based on pixels that belong to spectrally similar regions, which is identified using a spectral similarity measure (Park and Kang 2004).

4.5.2 Regression Variable Substitution

Multiple regression derives a variable, as a linear function of multivariable data, that will have maximum correlation with univariate data. In image fusion, the regression procedure is used to determine a linear combination (replacement vector) of an image channel than can be replaced by another image channel. This method is called regression variable substitution (RVS) (Pohl and Genderen 1998). To achieve the effect of fusion, the replacement vector should account for a significant amount of variance or information in the original multivariate data set. It is a pansharpening approach and a method for change detection using the assumption that pixels acquired at time one are a linear function of another set of pixels received at time two. Using the predicted value obtained from the least-squares regression, the difference image is the regression value—pixel of time one (Singh 1989).

4.5.3 Principal Component Analysis

PCA-based fusion follows a statistical approach. The techniques that follow component substitution using PCA have already been described in Section 4.3.2. They are the most common implementations of PCA in image fusion. However, there is also a part that uses PCA on all input bands of multi-sensor data that is described in this section. For PCA fusion, the image channels of the different sensors are combined into one image file and a PCA is calculated from all the channels.

Two types of PCA can be performed: *selective* or *standard*. The latter uses all available bands of the input image, for example, Landsat TM bands 1–7, the selective PCA uses only a selection of bands which are chosen based on *a priori* knowledge or application purposes, for example, would exclude the thermal infrared channel of TM (band 6). In case of Landsat 4 and 5, the first three PCs contain 98%–99% of the variance and therefore are mostly sufficient to represent the information.

The resulting principal components are presented in an RGB overlay for visual interpretation. Some examples of image fusion applying both PCS and PCA in comparison are reported by Yésou et al. (1993a) in the context of geological applications.

A similar approach to the PCA is accomplished in the C-stretch (color stretch) (Rothery and Francis 1987) and the D-stretch (de-correlation stretch) (Ehlers 1987). The de-correlation stretch helps to overcome the perceived problem that the original data often occupy a relatively small portion of the overall data space. In D-stretching, three-channel multispectral data are transformed onto principal component axes, stretched to give the data a spherical distribution in feature space and then transformed back onto the original axes. In C-stretching, PC_1 is discarded, or set to a uniform DN across the entire image, before applying the inverse transformation. This yields three color-stretched bands, which, when composited, retain the color relations of the original color composite but albedo and topographically induced brightness variations are removed.

4.5.4 Minimum Mean-Square Error Approach

This fusion approach is limited to pansharpening and based on the injection model to introduce high-resolution spatial detail in multispectral imagery. In the intention to develop an optimum pansharpening algorithm that responds to quality factors of the fused image, the MMSE fusion was developed. This method has been designed to optimize the parameters involved in the high spatial detail injection in wavelet fusion. The aim is to empirically improve spatial enhancement for very high-resolution data using low-pass filters of which the frequency responses match the shape of the MTF. The idea behind developing such an algorithm is to consider the effect of different parameters and processes on the quality of the fused image in the process of pansharpening in the context of injecting spatial detail into multispectral data. The algorithm integrates a linear injection model and the local MMSE parameter estimation (Garzelli et al. 2008). The processing flow is illustrated in Figure 4.13.

4.5.5 FuzeGo™

A very interesting and successful image fusion technique has been developed at the University of New Brunswick in Canada. The algorithm called UNB Pansharp has been commercialized and established in common image processing software packages (Zhang and Mishra 2014). It is based on least squares to best approximate the gray value relationship between the original MS and PAN to obtain a fused image with an optimized quality in spectral and spatial content. For best results, the input reference image channels have to be selected in such a way that the multispectral bands cover as close as possible the frequency range of the high-resolution panchromatic image, which is valid for most image fusion techniques as previously pointed out. In the meantime, the UNB algorithm has been converted into individual stand-alone software operating with regression analysis to produce weight factors, creating a synthetic image by multiplication and addition. The adaptive and context-specific approach preserves spectral integrity while adding spatial detail. Owing to the fact that the software is run commercially, the detailed algorithm behind it is hidden. Therefore, it is not possible to give a more detailed description.

4.5.6 Regression Fusion

Regression fusion (RF) uses the existing correlation between the lower resolution multispectral image and the high-resolution panchromatic data to

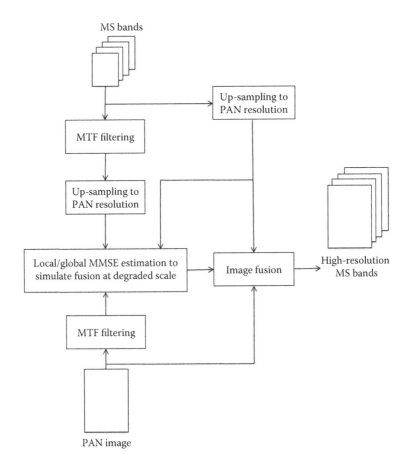

FIGURE 4.13
MMSE pansharpening process flowchart. (Adapted from Garzelli, A. et al. 2008. *IEEE Transactions on Geoscience and Remote Sensing*, 46 (1): 228–236.)

establish a relationship to fuse the image bands. Bias a_i and scaling b_i are calculated using the least-square approach from the resampled multispectral band and the high-resolution channel. The fused image is created using Equation 4.47:

$$XS_f = a_i + b_i * PAN \tag{4.45}$$

Using a sliding window, the method can be improved because local variations in correlation are considered. However, the success is limited to correlated bands only (Bretschneider and Kao 2000).

4.6 Multiresolution Approaches

In the literature, the techniques contained in the category of multiresolution approaches (MRA) (Nunez et al. 1999) carry many different names, which make their identification difficult and is confusing to newcomers in the field. The reason for this confusion is the fact that algorithms are identified either based on filtering scheme or on the form of injection. In an attempt to provide a generic framework for MRA pansharpening, Vivone et al. (2015) provide a list of methods describing terminology along with MRA schemes, filters, and gain used (see Table 4.2). The table illustrates categorization and terminology development using six examples.

Other common terms in the context of MRA are "Amélioration de la Résolution Spatiale par Injection de Structures" (ARSIS) (Wald and Ranchin 2003), that is, improving spatial resolution by structure injection, multiscale transform (MST) (Tai and Nipanikar 2015), multiscale decompositions (MSD), multiresolution wavelet decomposition (MWD) (Fanelli et al. 2001; Chibani and Houacine 2002), bilateral decomposition (Jianwen and Shutao 2011), pyramid transform (PT), and HFI (Schowengerdt 1980; Rong et al. 2014). Often, authors only mention the name of the algorithm used in the MRA fusion, that is, wavelet transform (WT), discrete wavelet transform (DWT), additive wavelet transform (AWT), contourlet transform, curvelet transform, and others. This category inherits some complexity not least because of the diversity of naming. The abbreviations appearing in the literature require careful

TABLE 4.2

Selected MRA Terminology Based on Decomposition, Filters, and Injection

Method	MRA	Filter	Gain	References
HPF	ATWT	Box filter	1	Chavez et al. (1991)
HPM	ATWT	Box filter	$\dfrac{MS_{res}}{PAN_{LP}}$	Schowengerdt (2007b)
SFIM	ATWT	Box filter	$\dfrac{MS_{res}}{PAN_{LP}}$	Liu (2000)
MTF-GLP	GLP	MTF	1	Aiazzi et al. (2006)
MTF-GLP-CBD	GLP	MTF	$\dfrac{cov(PAN_{LP}, MS_{res})}{var(P_{LP})}$	Aiazzi et al. (2006)
ATWT	ATWT	S&M	1	Vivone et al. (2014)

Source: Vivone, G. et al. 2015. *IEEE Transactions on Geoscience and Remote Sensing*, 53(5): 2565–2586.

HPM = high-pass modulation.

MS_{res} is the multispectral image at the scale of the panchromatic image.

PAN_{LP} is the low-pass version of the panchromatic image.

CDF = Cohen–Daubechies–Fauveau.

S&M = Starck and Murtagh.

reading in order to identify similarities and differentiate separate algorithms. Examples are MRA versus MSD (multiresolution approach vs. multiscale decomposition) or ATWT versus ATWD (à-trous wavelet transform vs. à-trous wavelet decomposition) (Chibani 2006). There are many names for the same approach, for example, à-trous wavelet transform (ATWT), undecimated discrete wavelet transform (UDWT), and DWT. Other names for this WT are stationary wavelet transform, shift invariant wavelet transform, and redundant wavelet transform. This makes an understanding and a comparison rather difficult.

Acknowledged problems of MRA in image fusion result from the necessary filtering process that may create ringing artifacts in the fused image (Garzelli et al. 2008).

4.6.1 Principles

With increase in computational power and availability of algorithms in commercial remote sensing software, the powerful multiresolution analysis techniques, such as wavelets, curvelets, and others, have become popular in recent years. MRA-based approaches decompose images into multiple channels depending on their local frequency content. The pyramid is used to represent the multiscale models of which the basis is the original image as shown in Figure 4.14. With increasing level, the original image is approximated at coarser spatial resolution. At each level of subsampling of the image, the spatial resolution is halved. Each input image results in four planes at each resolution level corresponding to one approximation image. There is one coarse spatial resolution image called low–low (LL), and there are three detailed images, that is, horizontal (low–high LH containing horizontal edge information), vertical (high–low HL with vertical edge information), and diagonal (high–high HH comprising diagonal edge information) (Bao and Zhu 2015). An MRA with N levels has $M = 3N + 1$ frequency bands.

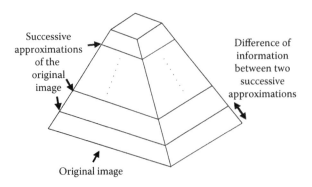

FIGURE 4.14
Pyramid processing in MRA fusion techniques.

The WT coefficients describe the local variation in the pixel neighborhood through scales. Therefore, the coefficient value for a particular location at any scale is equivalent to a measure of importance of a detail, meaning the higher the WT coefficients' amplitudes, the more important is the detail.

In practice, MRA consists of four steps (Vivone et al. 2015):

1. Resampling of the MS to the spatial resolution of the PAN.
2. Synthesizing of the PAN image.
3. Calculation of the band-dependent injection gains.
4. Perform the injection resulting in the fused image.

4.6.2 Fusion Process

The transform between the individual pyramid levels is performed using WT, Laplacian pyramid (LP), Gaussian pyramid (GP), curvelets, and others. In image fusion, the WT approach uses the following:

- *Substitution:* Selected MS wavelet planes are substituted by the planes of the corresponding PAN image
- *Addition:* Decomposed PAN planes are added to the MS bands or to MS intensity (Nunez et al. 1999)

The additive wavelet method follows four major steps as depicted in Figure 4.15. After having created a histogram-matched PAN* for each multispectral band i of n number of bands (1) the wavelet transform is applied to all resulted bands (2). The obtained detail images from the transformed PAN* are added to the corresponding transformed MS_i (3). Finally, the inverse WT returns the n fused bands that lead to the fused high-resolution multispectral image (4).

Fusion based on MRA requires a model to describe how high-pass information from PAN is injected into resampled MS bands. Basic WT substitution methods using the discrete WT (DWT) fuse sub-bands of corresponding frequencies, that is, LL, LH, HL, and HH. These decompositions exist at several levels, which form the pyramid. The fused image is produced by the inverse transform. In practice, the scaling and the WT function do not have to be explicitly derived. They are described by coefficients, which are fused by different fusion rules to produce the final image.

4.6.3 Advances in MRA Fusion

The additive wavelet (AWL) method was originally designed for three-band MS images. The PAN structure is injected into the luminance band of the original lower-resolution MS image (Nunez et al. 1999). It was extended to suit any number of bands resulting in the proportional AWL (AWLP) by

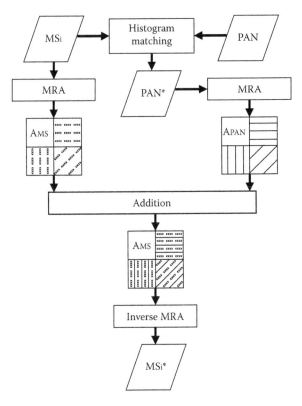

FIGURE 4.15
Image fusion process following an MRA.

Otazu et al. (2005). Better results can be obtained when the fusion process is context-driven, like the context-based decision (CBD) fusion model. This relates to how the high-pass information (spatial detail) is injected, in this case depending on the context (local instead of global). The local gain factor is given by the ratio of standard deviations of the fused MS band to the PAN. Particularly suitable are the generalized LP (GLP), the undecimated discrete wavelet transform (UDWT), or the popular ATWT (Aiazzi et al. 2002).

Recently, contourlets found their way into fusion algorithms. Contourlet transform (CT) "... capture and link discontinuity points into linear structures (contours) ..." (Metwalli et al. 2014). Their strength is the ability to have different number of directions at each scale of multiresolution decomposition. There are two stages to derive the contourlet coefficients, namely, a multiscale transform using a Laplacian pyramid (Alparone et al. 1998) to identify discontinuities and a local directional transform to group the wavelet-like coefficients to form a smooth contour. Contourlets allow flexible number of orientations (Amro et al. 2011). The nonsubsampled CT (NSCT) works on a nonsubsampled pyramid (NSP) and was found to produce better results in image fusion. It is a technique that can be found as a hybrid component together with PCA and IHS (Shah et al. 2008; Yang and Jiao 2008).

4.7 Hybrid Techniques

Different image fusion techniques lead to different achievements along with different limitations. To overcome the limitations and take advantage of the benefits of the different approaches, researchers have developed hybrid fusion techniques, combining two or more fusion algorithms in one workflow. In the following sections, we explain successful implementations of various fusion methods in hybrid techniques.

4.7.1 Ehlers Fusion

The spectral incompatibility of PAN and MS occupied many researchers. One research team tackled the problem by excluding the insertion of gray values to the spectral components of MS while increasing spatial detail using the PAN channel. An IHS transform converts the original MS bands into IHS space. A subsequently applied FFT transforms the intensity component and the PAN image. In the frequency domain, the intensity spectrum is low-pass filtered, and the PAN spectrum is high-pass filtered. The enhanced PAN spectrum is added to the multispectral spectra. With an inverse FFT, the images are converted back into the spatial domain. The actual fusion uses IHS using the computed intensity and an inverse IHS to fuse the data as shown in Figure 4.16. Owing to the many processing steps, we have split the workflow into two parts (a) and (b). This hybrid method is called Ehlers fusion. It overcomes the limitations of other methods even for multi-sensor or multi-temporal images (Ehlers 2004; Klonus 2008).

4.7.2 IHS–BT Fusion

The idea to provide adjustable parameters in image fusion approaches helps to optimize spectral and spatial content of fused images. Combining two popular algorithms, namely, IHS and BT, Su et al. (2013) propose the IHS–BT that allows parameter adjustment to represent either a pure IHS or a BT, or apply hybrid IHS–BT fusion. Additionally, it is possible to weight the contribution of SAR and PAN to optimize VIR/SAR fusion results. They overcome the issues of IHS and BT in the distortion of spectral information by balancing saturation using a parameter to control the degree of saturation stretch (BT) and compression (IHS).

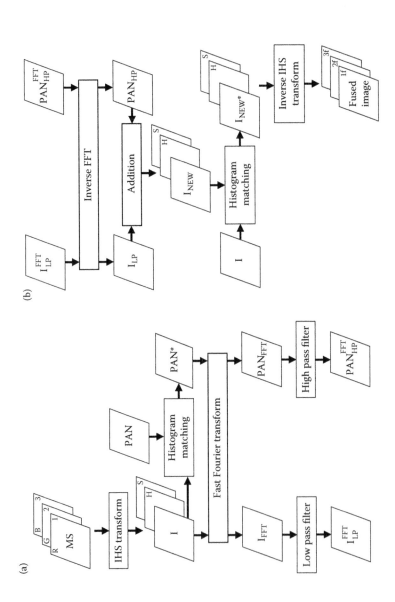

FIGURE 4.16
Ehlers fusion presented in two parts: (a) extracts the information from the multispectral and panchromatic image in the frequency domain; (b) shows the actual fusion process.

Mathematically, they presented the Brovey and IHS transforms in a generic algorithm in order to pinpoint the major effects of the two algorithms. As described in the beginning of this chapter, the RGB to IHS conversion model can follow a linear or a nonlinear transformation, as shown in Equations 4.4 through 4.7 and 4.9 through 4.11, respectively. According to their assessment, they found that the IHS method performs best using the nonlinear equations, while for BT the linear equations lead to the least distortions, especially for the saturation values. The algorithm they derived from these recognitions is listed as Equation 4.46:

$$\begin{pmatrix} R'_{\text{IHS-BT}} \\ G'_{\text{IHS-BT}} \\ B'_{\text{IHS-BT}} \end{pmatrix} = \frac{\text{PAN}}{I + k(\text{PAN} - I)} \begin{pmatrix} R + k(\text{PAN} - I) \\ G + k(\text{PAN} - I) \\ B + k(\text{PAN} - I) \end{pmatrix} \qquad (4.46)$$

where the $R'_{\text{IHS-BT}}$, $G'_{\text{IHS-BT}}$, and $B'_{\text{IHS-BT}}$ represent the three fused channels for red, green, and blue, respectively. k is a weighting factor with a value between 0 and 1. It controls the degree of saturation stretch (Su et al. 2013).

4.7.3 Wavelet-IHS Fusion

Other researchers tried to overcome the trade-off between spatial and spectral optimization by combining the IHS approach with MRA techniques, such as WT (Chibani and Houacine 2002; Gonzalez-Audicana et al. 2004; Zhang and Hong 2005). For the wavelet IHS (WIHS), the multispectral data are transformed by IHS of which the intensity I component is decomposed using WT to reach the same pixel size as PAN. The PAN channel is histogram matched to I and also WT decomposed. The I wavelet decompositions are replaced by the PAN decompositions, using weighted combination (Zhang and Hong 2005). The process is displayed in Figure 4.17.

4.7.4 Wavelet-PCA Fusion

MRA to which wavelets belong allows the injection of spatial detail using the high-frequency coefficients. If this is added to a principal component substitution (similar to WT-IHS fusion), the spatial detail of the PAN channel obtained by the MWD can be inserted into PC_1 using the inverse MWD process and the detail coefficients. An adjustment of the histogram in the form of matching is needed prior to the fusion process. Again the prerequisite is that both input images are coregistered. The multispectral image MS is supposed to be resampled to the pixel resolution of the panchromatic data (PAN). This hybrid method works similarly with PCA and IHS and was first introduced by Gonzalez-Audicana et al. (2004).

Following the flow in Figure 4.18, the up-sampled MS is entered into the PCA process to convert the channels into PCs. PAN is histogram matched to

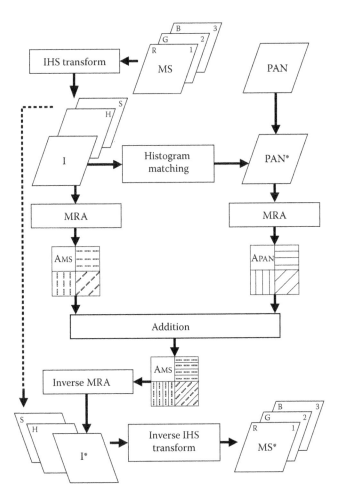

FIGURE 4.17
Hybrid fusion process using WT-IHS.

PC_1. The wavelet transform is applied to PC_1 and to the histogram-matched PAN*. Each MRA process delivers the four planes at each resolution level (LL, LH, HL, and HH) and the wavelet coefficient zero-mean images. The latter contain the spatial detail information. The spatial detail injection process takes place during the inverse wavelet transform where PC_1 receives the detail, for example, through addition. Only then the inverse PCA converts the data back into its original space resulting in the multispectral high-resolution fused image.

4.7.5 Modified BT-Wavelet Fusion

Chibani (2007) combined the modified Brovey transform (MBT) with ATWD to integrate PAN and SAR features with MS information. Figure 4.19 shows

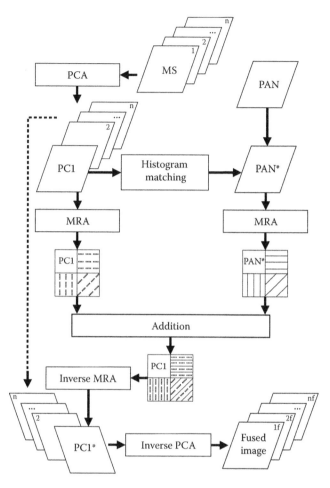

FIGURE 4.18
Hybrid fusion process using WT-PCA.

the workflow in which *I** is the resulting intensity from the IHS-MRA approach presented in Figure 4.17 and described in Section 4.7.3. The advantage of this approach is the possibility of producing the fused MS image without going through IHS transform. In addition, ATWD supports the extraction of spatial detail from PAN and SAR images to be injected as features into the MS image.

4.7.6 Laplacian Filtering and Multiple Regression

The essential advancement is the use of parameters for spectral/spatial quality adjustment. For pansharpening, the development of general frameworks containing optimization parameters has resulted in a general hybrid algorithm based on the work by Aiazzi et al. (2009). Choi et al. (2013) use

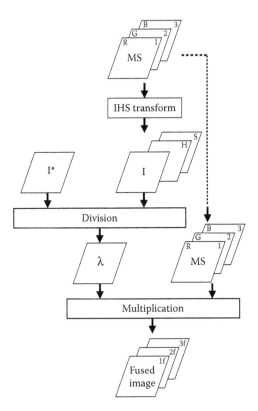

FIGURE 4.19
Hybrid fusion process using the modified BT and WT.

their equation and add Laplacian filtering along with an adjusted intensity image using multiple regression between MS and histogram-matched PAN. Their adjustment parameters use entropy and sensor-specific gain derived empirically.

It becomes obvious that optimum results incorporate image-dependent parameters. The success of hybrid algorithms lies in the fact that they are capable of accommodating different aspects of image content and quality adjustable to user needs. Each individual approach has disadvantages. Hybrid algorithms allow the combination of benefits from either of the input techniques.

4.8 Other Fusion Techniques

This group contains a description of the Bayesian data fusion (BDF), the spatial and temporal adaptive reflectance fusion model (STARFM), its derivative enhanced STARFM (ESTARFM), and the spatial temporal adaptive algorithm for mapping reflectance change (STAARCH). Even though not really considered a fusion technique classification is introduced in this section because it also efficiently exploits jointly processed multi-sensor data.

4.8.1 Classification

The principle of classification refers to the process of assigning each image pixel to a class defined by the user. The class corresponds to a certain type of land cover or use. Obviously, the classification of multimodal images leads to improved discrimination capability for the various classes if complementary data are used. Different sensor data require different preprocessing to account for sensor-specific distortions and variability as discussed previously in Chapter 3. A common approach to tackle multiple data source in an integrated fashion is the generation of multi-sensor/multisource band stacks, which is then fed into a classification. This is not an image fusion per se but protects the originality of the input and produces satisfactory results with minimum alteration and effort. In the literature, the term "layer stack" (originating from a function in ERDAS) or "stacked vector" (Simone et al. 2002) are common. According to Gomez-Chova et al. (2015), fusion in classification can be performed at feature, classifier, or output level. Those fusion techniques are not considered in this section since we concentrate on RSIF.

Popular pixel-based classifiers are

- Maximum likelihood classifier (MLC)
- Support vector machine (SVM)
- Classification tree analysis (CTA)

MLC is based on the assumption that a land cover type (class) is represented in the image with a certain range of values that create among the different image bands of a spectral signature. It requires the spectral response of each class to follow a Gaussian distribution. The parameters are calculated from selected training samples taken by the operator. In the classification process, their probability is maximized to assign each pixel to a certain class (Colditz et al. 2006). The results of MLC are limited in quality due to the fact that most data do not follow a Gaussian distribution and the possibility of class separation is reduced. SVM is a nonparametric or discriminative, binary classifier, which does not rely on distribution. It is based on statistical learning theory. Its advantage lies in its good performance even if only very few training samples are available. CTA is based on decision tree learning, a method commonly used in data mining. A decision tree is defined as a tree representing "…a predictive model expressed as a recursive partition of the covariates space to subspaces that constitute a basis for prediction" (Rokach 2016). The goal is to create a model that predicts the value of a target variable based on several input variables. The tree has to be trained using training samples for all classes. It is called tree because the classification starts with a very generic class separation (root node), for example, water and non-water, which is then split again and again (children nodes) into more definite subclasses until the desired classes are reached (leaves—final partitions). Known advantages of CTA lay in the fact that they form a meaningful classification and provide the option to convert into a set of *if–then* rules. Decision trees are independent of space distributions and can deal with diverse and large input data. CTA is robust and performs well even if only a few training samples exist. It is worth mentioning that fuzzy classifiers gain popularity because the traditional hard classifiers are not capable of handling detailed structures in very high-resolution data (Kuria et al. 2014).

Variations of this idea are pansharpening prior to classification or the use of ratios, that is, vegetation and other indices in combination with classification. The impact of pansharpening on classification has been studied in the past leading to the fact that pansharpening can increase classification accuracy but that it largely depends on the chosen algorithm (Amarsaikhan and Douglas 2004; Colditz et al. 2006; Amarsaikhan et al. 2012; Palsson et al. 2012; Johnson et al. 2014; Pohl and Hashim 2014). Owing to the availability of more computer power, higher spatial resolution imagery and the desire to understand complex Earth dynamics, OBIA gained importance. In OBIA, the algorithms are performed on features (objects) previously extracted from the original image. Feature extraction is done by segmentation. OBIA are superior to pixel-based processing because the extracted objects are a more intelligent representation of the real world, which again lead to more meaningful statistics in the process (Colditz et al. 2006; Jing and Cheng 2011; Johnson et al. 2013; Aguilar et al. 2014). Applications of multimodal classification are manifold, including land cover mapping (Farah et al. 2008), agriculture (Johnson et al. 2014), forestry (Deng et al. 2014; Ghulam 2014), urbanization (Amarsaikhan et al. 2013; Idol et al. 2015), change detection (Stefanski et al. 2014) and others (Gomez-Chova et al. 2015; Zhang 2015) (see also Chapter 6).

4.8.2 Spatial and Temporal Data Fusion Approaches

For the integration of very different spatial resolution optical remote sensing images, spatial and temporal data fusion approaches (STDFA) have evolved. Within this category, STARFM is a particular model developed to accommodate high temporal and high spatial resolution images from Landsat and MODIS (Feng et al. 2006). The input images have to be calibrated and atmospherically corrected to surface reflectance values. STARFM fuses images that are strongly correlated at spectral and spatial levels. This requires an overlapping of spectral bands to be successful. Physical attributes at pixel level are combined. The software can be downloaded from the Landsat ecosystem disturbance adaptive processing system (LEDAPS) website (http://ledapsweb.nascom.nasa.gov). The strength of STARFM is its good performance over homogeneous areas and its ability to work either one or two image pairs, which is an advantage in case of no data availability for Landsat (e.g., cloud cover).

The algorithm has been advanced into accommodating heterogeneous landscapes and other sensors than Landsat and MODIS. This improved algorithm is called enhanced STARFM (ESTARFM) (Zhu et al. 2010). Last but not least, STAARCH serves change detection of reflectance values due to land cover change and disturbance (Hilker et al. 2009; Zhu et al. 2010).

Where STARFM leads to less accurate results, the STAARCH algorithm detects changes in reflectance using Tasseled Cap transformations of the input images. Based on the date of disturbance (DoD—Landsat) and disturbance index (DI—MODIS), pixels are flagged as disturbed. Depending on the disturbance rating, the approach selects the most appropriate image pairs for surface reflectance prediction for the synthetic Landsat image (Hilker et al. 2009). STAARCH carries the advantage of being able to accommodate abrupt changes in the period of observations and the time period can be longer than for STARFM and ESTARFM (Gao et al. 2015).

Both STARFM and STAARCH have been developed to increase the temporal resolution of 16-day repeat-cycle Landsat using daily MODIS images. ESTARFM can be applied to optical remote sensing images other than MODIS and Landsat. It is capable of dealing with more heterogeneous areas because it introduces land cover-related conversion coefficient. All three approaches assume that changes in the surface reflectance at the two dates are consistent and comparable at the different resolutions. They consider that the image pixels change proportional to each other within each image pair. A model translates changes between the sensors based on spectral reflectance, considering neighboring pixels to predict the central pixel in the moving window (Gao et al. 2015).

For ESTARFM, two pairs of high/low spatial resolution images acquired at two different times (t_1, t_2) are needed. For the prediction date t_p, a low spatial resolution (LSR) image is necessary (Zhu et al. 2010). For this date, the high spatial resolution (HSR) image will be synthesized using ESTARFM. In total, there are four steps (Tewes et al. 2015):

1. In two HSR images, a moving search window, sized to one LSR image pixel, searches for similar pixels to the central pixels.
2. A weight w_i is generated based on the distance of the found pixel to the central pixel.
3. A linear regression is calculated using the LSR pixel values at the two times against the HSR similar pixels. This creates the conversion coefficients, which convert the change from the LSR to the HSR image.
4. Calculation of the synthesized HSR at the predicted date using the formula derived in step 3 using Equation 4.47.

$$F(x,y,t_p) = H(x,y,t_j) + \sum_{i=1}^{N} w_i \times v_i \times (L(x_i,y_i,t_p) - L(x_i,y_i,t_j)) \qquad (4.47)$$

where H and L stand for the reflectance of the HSR and LSR images, respectively, (x, y) is the location of the predicted HSR pixel value, and (x_i, y_i) is the location of the ith similar pixel. The j refers to the time of the input pair chosen (t_1 or t_1), while N represents the total number of similar pixels of the predicted HSR image in a chosen window (Tewes et al. 2015). Both STAARCH and ESTARFM only work with two image pairs.

In terms of algorithm development, there are still limitations in the use of ESTARFM. Like STARFM, the prediction of objects that change shape in time is blurred at the changing boundary. In addition, short-term changes that do not appear in the fine-resolution images are not predictable. A prerequisite for a successful performance is the need for a linear relationship of the spectral bands used and performance depends on a proper image selection in general (Zhu et al. 2010).

4.9 Selection Approach

Some data combinations require special considerations in terms of algorithm choice. Image pairs originating from optical and microwave sensors used to be exploited strictly separate. However, this has changed since the successful implementation of VIR/SAR image fusion in various applications around the globe. During the past 15 years, we have experienced a tremendous increase in new satellite launches in Earth observation. Users and researchers benefit from the growing availability and accessibility of remote sensing images from satellites at increasing spatial and spectral resolutions. This has given a boost to the development of multi-sensor image fusion approaches as well as to the exploitation of new applications. A special category is the fusion of complementary imagery from passive optical and active radar sensors. Special attention has to be paid to accurate geometric processing prior to fusion due to the sensors' different ground displacements (Palubinskas et al. 2010). The diagram shown in Figure 4.20 provides a good overview on the popularity of different sensors for image fusion. IKONOS, QuickBird, Landsat, and SPOT are still among the most commonly used image providers in image fusion. Sensors mentioned less than 10 times were grouped into "others." The latter contains published experiments with RapidEye, Zy-3, CosmoSkymed, Formosat, Kompsat, and UAV data. Recently, researchers have also concentrated on LiDAR/remote sensing image fusion (Berger et al. 2013; Zhou and Qiu 2015) and combining

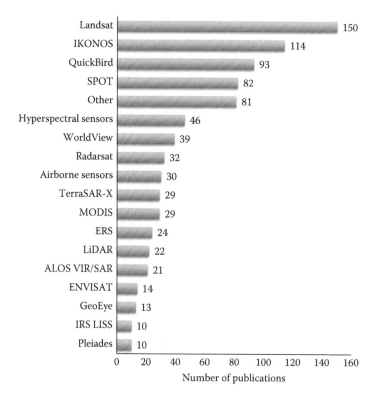

FIGURE 4.20
Popularity of sensors in image fusion experiments published between 1998 and 2015. (Adapted from Pohl, C. and J. L. van Genderen. 2015. *International Journal of Image and Data Fusion* 6 (1):3–21.)

satellite images with unmanned airborne vehicle (UAV) acquired images or multiple UAV acquired images (Cho et al. 2013; Delalieux et al. 2014; Gevaert 2014).

4.9.1 VIR/SAR

Active SAR and passive optical remote sensing produce data types that have complementary but very different characteristics and information content; the number of algorithms capable of coping with this is limited. The exploitation of this spatio-spectral data has become very popular because of higher reliability in information extraction and accessibility to operational, high-resolution SAR sensors. There will be another push in this respect since Sentinel-1 SAR is publically available at no costs. In the context of RSIF, a comparison of eight commonly used approaches to fuse TerraSAR-X with QuickBird imagery resulted in the identification of the Ehlers method as best solution in terms of simultaneous high spectral and spatial quality (Klonus 2008). This is confirmed by Berger (2010) who combined TerraSAR-X with RapidEye images. Both researchers consider WT approaches equally suitable with a minor drawback in spatial detail. However, HPF resolution merge also constantly performed well in VIR/SAR fusion, no matter the quality aspect evaluated. PCA, CN, GS, or UNB are suitable for single-sensor, single-date images. They fail when multi-sensor images such as VIR/SAR combinations are involved (Ehlers et al. 2010; Abdikan et al. 2012).

Zhang et al. (2010) suggested using linear regression of block regions to combine VIR and SAR images. This has the advantage of optimizing the trade-off between spectral and spatial information depending on the local image characteristics. In addition, the block processing decreases computational requirements. Others employed the ensemble empirical mode decomposition (EMD) to fuse ATWT extracted SAR features with the MS image reducing the amount of spectral distortion compared to HPF, ATWT, and IHS methods (Chen et al. 2011). Other researchers solve the data diversity problem through intensity modulation of the MS image by integrating SAR texture after pansharpening (Alparone et al. 2004). After having applied a speckle filter on the SAR image, the ratio of the image and its low-pass approximation using an ATWT is computed. The latter contains the modulating texture that is induced into the fusion process that basically applies the modified IHS.

4.9.2 Feature Extraction

Advanced fusion approaches use feature extraction to selectively induce spatial detail in a PAN image. The modified PAN is then fused with MS data (Byun et al. 2013). This innovative procedure is adaptive and selects the fusion rule depending on homogeneity or heterogeneity of the SAR texture information. Following the MRA/IHS-based approach, it was suggested to use weights in the process of SAR feature injection into the intensity component of MS images (Hong et al. 2009). Summarizing the efforts in VIR/SAR image fusion, successful fusion algorithms account for the spectral differences, such as the Ehlers method, use preprocessing (feature extraction for SAR), or are applied context-driven (choosing different fusion rules depending on the homogeneity of the regions).

4.9.3 Optimum Index Factor

Last but not least, we would like to mention that in the beginning of remote sensing image fusion research in the early 1980s, the limitation of fusion techniques to three input bands led to the development of an interesting band selection criteria. One method originally designed to select the most appropriate Landsat MSS bands to create useful band ratios for interpretation relies on statistics in order to select the data containing most of the variance. As mentioned in Section 3.1, this is the selection method developed by Chavez et al. (1982) called *optimum index factor* (OIF) mathematically described in Equation 3.1.

Other dimensionality reduction methods, such as PCA, MNF, HySime, and ISOMAP (Selva et al. 2015), play an important role in hyperspectral remote sensing in particular and has been discussed in Chapter 1. The optimum number of bands can be determined by a minimum number of spectrally distinct signal sources that define the hyperspectral data. This concept is called virtual dimensionality (VD) and uses eigenvalues (Chein and Qian 2004).

4.10 Comparison and Popularity of Fusion Techniques

The most interesting aspects in studying the achievements of research in remote sensing image fusion are the comparison of communalities and contradictions from the various experiments (Pohl and Genderen 2015). There is no such thing as the "best fusion method" because the choice very much depends on the task (application, desired information, interpretation method), size/time constraints (speed, complexity, knowledge), as well as geographic location, budget, and more. The study of published research outcomes in international, peer-reviewed journals led to an overview on RSIF techniques popularity. The outcome is presented in Figure 4.21.

There have been many efforts to provide a complete overview on existing remote sensing image fusion techniques (Pohl and Genderen 1998; Jinghui et al. 2010; Khaleghi et al. 2013; Pohl and Genderen 2014). It remains an important task due to ongoing development of new sensors and new algorithms. The most common RSIF algorithms according to the literature are CS, MRA, and lately, hybrid algorithms along after arithmetic methods as shown in Figure 4.21. CS is very popular, mostly because of its simplicity and straightforwardness. MRA has caught up popularity ranking because of the availability and accessibility of more powerful computing facilities. It has shown to produce good quality. Hybrid algorithms are a major trend since they allow benefiting from the advantages of different algorithms and get rid of trade-offs that exist if only one algorithm is applied. If we include classification as fusion approach, it has become the third most accepted method to derive information from multi-sensor data.

In terms of performance, authors tend to achieve best results for their own proposed algorithms. When comparing the different results, the "best" algorithm changes from paper to paper. An attempt to compare the performance of different algorithms might be consistent for a certain experiment or publication. However, it is not feasible to assess the performance of fusion algorithms across experiments due to numerous incompatible factors

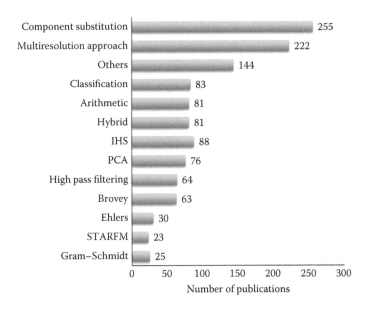

FIGURE 4.21
Most commonly applied RSIF techniques and algorithms published between 1998 and 2015. (Adapted from Pohl, C. and Y. Zeng. 2015. Development of a fusion approach selection tool. Paper read at *International Workshop on Image and Data Fusion*, 21–23 July 2015, Kona, Hawaii.)

in the various studies. The images used in the particular research project are different from study to study. Even if the same sensor source is used the covered area contains completely different characteristics. Each author selects a different set of algorithms to be compared. Even if the algorithm is identified and similar between two studies, the individual parameter choice within the algorithm can be different. Quality assessment is also biased since each time a different set of indices is applied.

A comparison attempt of four recent publications of international journals shows that the definition of the "best" technique remains an individual choice:

1. Xu et al. (2015) compared fused IKONOS images using GS, UNB, context-driven fusion (CDF), and enhanced context-based model (ECM) with their proposed adaptive pansharpening based on linear regression and histogram adjustment. They used Q4 (contrast, mean, bias, similarity), spectral angle mapper (SAM), *Error Relative Globale Adimensionnelle de Synthése* (ERGAS), and SNR for a quantitative assessment. Their proposed algorithm showed best performance.

2. Zhou et al. (2014) implemented GIHS and various other IHS variations, GS as well as a proposed IHS algorithm with additional edge restoration and spectral modulation (ERSM) using QuickBird and

WorldView-2 imagery. Quality assessment was applied through D_λ (spectral distortion measure), D_s (spatial distortion measure), ERGAS, saturation deviation index (SDI), SAM, spatial frequency SF, universal image quality index (UIQI), and Q4. Again, the proposed algorithm performed best.

3. Xu et al. (2014) follow an interesting quality assessment approach looking at the impact of miss-registration, object size difference, and impact on spectral information. Comparing QuickBird pansharpened images using UNB, GS, adaptive GIHS, WT, and BT, they identified UNB as the best algorithm in robustness.

4. Palsson et al. (2014) suggest a sophisticated algorithm based on an observational model integrated in a linear combination between PAN and MS. They compared their algorithm in experiments on Pleiades and QuickBird imagery, evaluating the results with D_λ, D_s (QNR—quality without reference), ERGAS, SAM, and cross-correlation (CC). In comparison with other algorithms, such as P + XS, PCA, and un-decimated WT (UDWT), their algorithm received best quality values.

The list of comparison papers is long. The value of intercomparison is doubtful since there is no common background and definitions to provide meaningful values. Even extracting a trend using equivalent comparable parameters between different publications is impossible because the different choices of images, techniques, covered area, and assessment criteria.

4.11 Summary

After a brief introduction of some issues in remote sensing image fusion, this chapter first of all has sorted the many image fusion algorithms, and classified them into six categories. In the literature, the many image fusion algorithms are given different names and use different descriptions. This chapter intended to clarify the confusion and organize the topic so that it can be well understood. The main part of the chapter then has gone into detail to describe each of the fusion categories in turn and describes the many image fusion algorithms used in remote sensing. Besides the details on each of the fusion techniques, the authors provide more than 40 computer formulae to show how each algorithm actually works. After having described each of the fusion techniques, the chapter discusses how to select the most appropriate method, and discusses the aspects to take into consideration in this selection process. The conclusion is that there is no one "best" method to select the most appropriate fusion algorithm, as this still depends on the application being dealt with, and on the user's objectives for fusing the imagery. Prior to presenting the many useful applications of RSIF in Chapter 6, it is important to discuss first the quality assessment of the fusion result, as one needs to know the reliability and accuracy of the fused result, before applying it to solve a particular application problem. The topic of accuracy assessment is discussed in the next chapter.

References

Abdikan, S., F. Balik Sanli, F. Sunar, and M. Ehlers. 2012. A comparative data-fusion analysis of multi-sensor satellite images. *International Journal of Digital Earth* 7 (8):671–687.

Aguilar, M., F. Bianconi, F. Aguilar, and I. Fernández. 2014. Object-based greenhouse classification from GeoEye-1 and WorldView-2 stereo imagery. *Remote Sensing* 6 (5):3554–3582.

Aiazzi, B., L. Alparone, S. Baronti, and A. Garzelli. 2002. Context-driven fusion of high spatial and spectral resolution images based on oversampled multiresolution analysis. *IEEE Transactions on Geoscience and Remote Sensing* 40 (10):2300–2312.

Aiazzi, B., L. Alparone, S. Baronti, A. Garzelli, and M. Selva. 2003. An MTF-based spectral distortion minimizing model for pan-sharpening of very high resolution multispectral images of urban areas. Paper read at *2nd GRSS/ISPRS Joint Workshop on Remote Sensing and Data Fusion over Urban Areas, 2003*, 22–23 May 2003, Berlin, Germany.

Aiazzi, B., L. Alparone, S. Baronti, A. Garzelli, and M. Selva. 2006. MTF-tailored multiscale fusion of high-resolution MS and PAN imagery. *Photogrammetric Engineering & Remote Sensing* 72 (5):591–596.

Aiazzi, B., L. Alparone, S. Baronti, A. Garzelli, and M. Selva. 2011. 25 years of pan-sharpening: A critical review and new developments. In *Signal and Image Processing for Remote Sensing,* 2nd edition, edited by C. H. Chen, Boca Raton, FL: CRC Press, 533–548.

Aiazzi, B., L. Alparone, S. Baronti, A. Garzelli, and M. Selva. 2012. Twenty-five years of pansharpening: A critical review and new developments. In *Signal and Image Processing for Remote Sensing*, edited by C. H. Chen. Boca Raton, FL: CRC Press.

Aiazzi, B., S. Baronti, F. Lotti, and M. Selva. 2009. A comparison between global and context-adaptive pansharpening of multispectral images. *IEEE Geoscience and Remote Sensing Letters* 6 (2):302–306.

Al-Wassai, F. A., N. Kalyankar, and A. A. Al-Zuky. 2011. The IHS transformations based image fusion. *Computer Vision and Pattern Recognition* 2 (5):1–10.

Alparone, L., S. Baronti, A. Garzelli, and F. Nencini. 2004. Landsat ETM+ and SAR image fusion based on generalized intensity modulation. *IEEE Transactions on Geoscience and Remote Sensing* 42 (12):2832–2839.

Alparone, L., V. Cappellini, L. Mortelli, B. Aiazzi, S. Baronti, and R. Carlà. 1998. A pyramid-based approach to multisensor image data fusion with preservation of spectral signatures. In *Future Trends in Remote Sensing*, edited by P. Gudmandsen, Rotterdam, The Netherlands: Balkema, 418–426.

Amarsaikhan, D., T. Bat-Erdene, M. Ganzorig, and B. Nergui. 2013. Knowledge-based classification of quickbird image of Ulaanbaatar City, Mongolia. *American Journal of Signal Processing* 3 (3):71–77.

Amarsaikhan, D. and T. Douglas. 2004. Data fusion and multisource image classification. *International Journal of Remote Sensing* 25 (17):3529–3539.

Amarsaikhan, D., M. Saandar, M. Ganzorig et al. 2012. Comparison of multisource image fusion methods and land cover classification. *International Journal of Remote Sensing* 33 (8):2532–2550.

Amro, I., J. Mateos, M. Vega, R. Molina, and A. Katsaggelos. 2011. A survey of classical methods and new trends in pansharpening of multispectral images. *EURASIP Journal on Advances in Signal Processing* 2011 (1):79.

Atkinson, P. M. 2013. Downscaling in remote sensing. *International Journal of Applied Earth Observation and Geoinformation* 22:106–114.

Bao, W. and X. Zhu. 2015. A novel remote sensing image fusion approach research based on HSV space and bi-orthogonal wavelet packet transform. *Journal of the Indian Society of Remote Sensing* 43 (3):467–473.

Berger, C. 2010. Fusion of high resolution SAR data and multispectral imagery at pixel level—A comprehensinve evaluation. MSc, Institute for Geography, Friedrich-Schiller-University Jena, Jena, Germany.

Berger, C., M. Voltersen, R. Eckardt et al. 2013. Multi-modal and multi-temporal data fusion: Outcome of the 2012 GRSS data fusion contest. *IEEE Journal of Selected Topics in Applied Earth Observations and Remote Sensing* 6 (3):1324–1340.

Bretschneider, T. and O. D. Kao. 2000. Image fusion in remote sensing. Paper read at *1st Onine Symposium of Electronics Engineers, OSEE 2000*, Waltham, USA.

Byun, Y., J. Choi, and Y. Han. 2013. An area-based image fusion scheme for the integration of SAR and optical satellite imagery. *IEEE Journal of Selected Topics in Applied Earth Observations and Remote Sensing* 6 (5):2212–2220.

Carper, W. J., T. M. Lillesand, and P. W. Kiefer. 1990. The use of intensity-hue-saturation transformations for merging SPOT panchromatic and multispectral image data. *Photogrammetric Engineering & Remote Sensing* 56 (4):459–467.

Chavez, P. S. and J. A. Bowell. 1988. Comparison of the spectral information content of Landsat Thematic Mapper and SPOT for three different sites in the Phoenix, Arizona region. *ISPRS Journal of Photogrammetry and Remote Sensing* 54 (12):1699–1708.

Chavez, P. S., S. C. Sides, and I. A. Anderson. 1991. Comparison of three different methods to merge multiresolution and multispectral data: Landsat TM and SPOT panchromatic. *Photogrammetric Engineering and Remote Sensing* 57 (3):295–303.

Chein, I. C. and D. Qian. 2004. Estimation of number of spectrally distinct signal sources in hyperspectral imagery. *IEEE Transactions on Geoscience and Remote Sensing* 42 (3):608–619.

Chen, S., H. Su, R. Zhang, J. Tian, and J. Xia. 2011. Scaling between Landsat-7 and SAR images based on ensemble empirical mode decomposition. *International Journal of Remote Sensing* 33 (3):826–835.

Chibani, Y. 2006. Additive integration of SAR features into multispectral SPOT images by means of the à trous wavelet decomposition. *ISPRS Journal of Photogrammetry and Remote Sensing* 60 (5):306–314.

Chibani, Y. 2007. Integration of panchromatic and SAR features into multispectral SPOT images using the "à trous" wavelet decomposition. *International Journal of Remote Sensing* 28 (10):2295–2307.

Chibani, Y. and A. Houacine. 2002. The joint use of IHS transform and redundant wavelet decomposition for fusing multispectral and panchromatic images. *International Journal of Remote Sensing* 23 (18):3821–3833.

Chiesa, C. C. and W. Tyler. 1990. Data fusion of off-nadir SPOT panchromatic images with other digital data sources. Paper read at *1990 ACSM-ASPRS Annual Convention*, Denver, CO.

Cho, K., E. Baltsavias, F. Remondino, U. Soergel, and H. Wakabayashi. 2013. Resilience against disasters using remote sensing and geoinformation technologies for rapid mapping and information dissemination (RAPIDMAP). Paper read at *34th Asian Conference on Remote Sensing 2013, ACRS 2013*, Bali, Indonesia.

Choi, J., J. Yeom, A. Chang, Y. Byun, and Y. Kim. 2013. Hybrid pansharpening algorithm for high spatial resolution satellite imagery to improve spatial quality. *IEEE Transactions on Geoscience and Remote Sensing Letters* 10 (3):490–494.

Choi, M. 2006. A new intensity-hue-saturation fusion approach to image fusion with a tradeoff parameter. *IEEE Transactions on Geoscience and Remote Sensing* 44 (6):1672–1682.

Cliche, G., F. Bonn, and P. Teillet. 1985. Integration of the SPOT pan channel into its multispectral mode for image sharpness enhancement. *Photogrammetric Engineering & Remote Sensing* 51:311–316.

Colditz, R. R., T. Wehrmann, M. Bachmann et al. 2006. Influence of image fusion approaches on classification accuracy: A case study. *International Journal of Remote Sensing* 27 (15):3311–3335.

Coppin, P., I. Jonckheere, K. Nackaerts, B. Muys, and E. Lambin. 2004. Review article. Digital change detection methods in ecosystem monitoring: A review. *International Journal of Remote Sensing* 25 (9):1565–1596.

Delalieux, S., P. J. Zarco-Tejada, L. Tits, M. A. J. Bello, D. S. Intrigliolo, and B. Somers. 2014. Unmixing-based fusion of hyperspatial and hyperspectral airborne imagery for early detection of vegetation stress. *IEEE Journal of Selected Topics in Applied Earth Observations and Remote Sensing* 7 (6):2571–2582.

Deng, S., M. Katoh, Q. Guan, N. Yin, and M. Li. 2014. Estimating forest aboveground biomass by combining ALOS PALSAR and WorldView-2 data: A case study at Purple Mountain National Park, Nanjing, China. *Remote Sensing* 6 (9):7878–7910.

Ehlers, M. 1987. Integrative Auswertung von digitalen Bildern aus der Satellitenphotogrammetrie und -fernerkundung im Rahmen von geographischen Informationssystemen. Dissertation, Vermessungswesen, Universität Hannover, Hannover.

Ehlers, M. 2004. Spectral characteristics preserving image fusion based on Fourier domain filtering. Paper read at *Remote Sensing for Environmental Monitoring, GIS Applications, and Geology IV*, Bellingham, WA, USA.

Ehlers, M., S. Klonus, P. Johan Åstrand, and P. Rosso. 2010. Multi-sensor image fusion for pansharpening in remote sensing. *International Journal of Image and Data Fusion* 1 (1):25–45.

Fanelli, A., A. Leo, and M. Ferri. 2001. Remote sensing images data fusion: A wavelet transform approach for urban analysis. Paper read at *IEEE/ISPRS Joint Workshop 2001, Remote Sensing and Data Fusion over Urban Areas*, Rome, Italy.

Farah, I. R., W. Boulila, K. Saheb Ettabaa, and M. Ben Ahmed. 2008. Multiapproach system based on fusion of multispectral images for land-cover classification. *IEEE Transactions on Geoscience and Remote Sensing* 46 (12):4153–4161.

Feng, G., J. Masek, M. Schwaller, and F. Hall. 2006. On the blending of the Landsat and MODIS surface reflectance: Predicting daily Landsat surface reflectance. *IEEE Transactions on Geoscience and Remote Sensing* 44 (8):2207–2218.

Franklin, S. E. and C. F. Blodgett. 1993. An example of satellite multisensor data fusion. *Computers & Geosciences* 19 (4):577–583.

Gangkofner, U. G., P. S. Pradhan, and D. W. Holcomb. 2008. Optimizing the high-pass filter addition technique for image fusion. *Photogrammetric Engineering & Remote Sensing* 747 (9):1107–1118.

Gao, F., T. Hilker, X. Zhu et al. 2015. Fusing landsat and MODIS data for vegetation monitoring. *IEEE Geoscience and Remote Sensing Magazine* 3 (3):47–60.

Garzelli, A. and F. Nencini. 2005. Interband structure modeling for Pan-sharpening of very high-resolution multispectral images. *Information Fusion* 6 (3):213–224.

Garzelli, A., F. Nencini, and L. Capobianco. 2008. Optimal *MMSE* Pan sharpening of very high resolution multispectral images. *IEEE Transactions on Geoscience and Remote Sensing* 46 (1):228–236.

Gevaert, C. 2014. Combing hyperspectral UAV and multispectral Formosat-2 imagery for precision agriculture applications, Department of Geography and Ecosystem Science, Lund University.

Ghulam, A. 2014. Monitoring tropical forest degradation in Betampona Nature Reserve, Madagascar using multisource remote sensing data fusion. *IEEE Journal of Selected Topics in Applied Earth Observations and Remote Sensing* 7 (12):4960–4971.

Gillespie, A. R., A. B. Kahle, and R. E. Walker. 1986. Color enhancement of highly correlated images. I. Decorrelation and HSI contrast stretches. *Remote Sensing of Environment* 20 (3):209–235.

Gomez-Chova, L., D. Tuia, G. Moser, and G. Camps-Valls. 2015. Multimodal classification of remote sensing images: A review and future directions. *Proceedings of the IEEE* 103 (9):1560–1584.

Gonzalez-Audicana, M., J. L. Saleta, R. G. Catalan, and R. Garcia. 2004. Fusion of multispectral and panchromatic images using improved IHS and PCA mergers based on wavelet decomposition. *IEEE Transactions on Geoscience and Remote Sensing* 42 (6):1291–1299.

Hallada, W. A. and S. Cox. 1983. Image sharpening for mixed spatial and spectral resolution satellite systems. Paper read at *17th International Symposium on Remote Sensing of the Environment*, Ann Arbor, Michigan.

Hilker, T., M. A. Wulder, N. C. Coops et al. 2009. A new data fusion model for high spatial- and temporal-resolution mapping of forest disturbance based on Landsat and MODIS. *Remote Sensing of Environment* 113 (8):1613–1627.

Hong, G., Y. Zhang, and B. Mercer. 2009. A wavelet and IHS integration method to fuse high resolution SAR with moderate resolution multispectral images. *Photogrammetric Engineering & Remote Sensing* 75 (10):1213–1223.

Idol, T., B. Haack, and R. Mahabir. 2015. Radar and optical remote sensing data evaluation and fusion; a case study for Washington, DC, USA. *International Journal of Image and Data Fusion* 6 (2):138–154.

Jawak, S. D. and A. J. Luis. 2013. A spectral index ratio-based Antarctic land-cover mapping using hyperspatial 8-band WorldView-2 imagery. *Polar Science* 7 (1):18–38.

Jianwen, H. and L. Shutao. 2011. Fusion of panchromatic and multispectral images using multiscale dual bilateral filter. Paper read at *18th International Conference on Image Processing (ICIP)*, Brussels, Belgium, September 11–14, 2011.

Jing, L. and Q. Cheng. 2011. An image fusion method based on object-oriented classification. *International Journal of Remote Sensing* 33 (8):2434–2450.

Jinghui, Y., J. Zhang, L. Haitao, S. Yushan, and P. Pengxiam. 2010. Pixel level fusion methods for remote sensing images: A current review. In *ISPRS TC VII Symposium—100 Years ISPRS*, edited by W. S. Wagner and B. Szekely. Vienna, Austria: ISPRS, pp. 680–686.

Johnson, B. A., H. Scheyvens, and B. R. Shivakoti. 2014. An ensemble pansharpening approach for finer-scale mapping of sugarcane with Landsat 8 imagery. *International Journal of Applied Earth Observation and Geoinformation* 33:218–225.

Johnson, B. A., R. Tateishi, and N. T. Hoan. 2013. A hybrid pansharpening approach and multiscale object-based image analysis for mapping diseased pine and oak trees. *International Journal of Remote Sensing* 34 (20):6969–6982.

Khan, M. M., J. Chanussot, L. Condat, and A. Montanvert. 2008. Indusion: Fusion of multispectral and panchromatic images using the induction scaling technique. *IEEE Geoscience and Remote Sensing Letters* 5 (1):98–102.

Khaleghi, B., A. Khamis, F. O. Karray, and S. N. Razavi. 2013. Multisensor data fusion: A review of the state-of-the-art. *Information Fusion* 14 (1):28–44.

Klonus, S. 2008. Comparison of pansharpening algorithms for combining radar and multispectral data. Paper read at *XXI ISPRS Congress*, Beijing, China.

Kuria, D. N., G. Menz, S. Misana et al. 2014. Seasonal vegetation changes in the Malinda wetland using bi-temporal, multi-sensor, very high resolution remote sensing data sets. *Advances in Remote Sensing* 3 (1):33–48.

Laben, C. A., V. Bernard, and W. Brower. 2000. Process for enhancing the spatial resolution of multispectral imagery using pan-sharpening. In http://www.google.com/patents/US6011875, edited by U.S. Patent. US & International: Laben et al.

Leonov, Y. P. and E. N. Sokolov. 2008. The representation of colors in spherical space. *Color Research & Application* 33 (2):113–124.

Leung, Y., J. Liu, and J. Zhang. 2014. An improved adaptive intensity-hue-saturation method for the fusion of remote sensing images. *IEEE Geoscience and Remote Sensing Letters* 11 (5):985–989.

Liu, J. G. 2000. Smoothing filter-based intensity modulation: A spectral preserve image fusion technique for improving spatial details. *International Journal of Remote Sensing* 21 (18):3461–3472.

Maurer, T. 2013. How to pan-sharpen images using the Gram-Schmidt pan-sharpen method—A recipe. *The International Archives of the Photogrammetry, Remote Sensing and Spatial Information Sciences* XL-1/W1:239–244.

Metwalli, M. R., A. H. Nasr, O. S. Faragallah et al. 2014. Efficient pan-sharpening of satellite images with the contourlet transform. *International Journal of Remote Sensing* 35 (5):1979–2002.

Moran, M. S. 1990. A window-based technique for combining Landsat Thematic Mapper thermal data with higher-resolution multispectral data over agricultural lands. *Photogrammetric Engineering & Remote Sensing* 56 (3):337–342.

Munechika, C. K., J. S. Warnick, C. Salvaggio, and J. R. Schott. 1993. Resolution enhancement of multispectral image data to improve classification accuracy. *Photogrammetric Engineering & Remote Sensing* 59 (1):67–72.

Nunez, J., X. Otazu, O. Fors, A. Prades, V. Pala, and R. Arbiol. 1999. Multiresolution-based image fusion with additive wavelet decomposition. *IEEE Transactions on Geoscience and Remote Sensing* 37 (3):1204–1211.

Otazu, X., M. Gonzalez-Audicana, O. Fors, and J. Nunez. 2005. Introduction of sensor spectral response into image fusion methods. Application to wavelet-based methods. *IEEE Transactions on Geoscience and Remote Sensing* 43 (10): 2376–2385.

Padwick, C., M. Deskevich, F. Pacifici, and S. Smallwood. 2010. Worldview-2 Pan-sharpening. Paper read at *ASPRS Annual Conference*, San Diego, California.

Palsson, F., J. R. Sveinsson, J. A. Benediktsson, and H. Aanaes. 2012. Classification of pansharpened urban satellite images. *IEEE Journal of Selected Topics in Applied Earth Observations and Remote Sensing* 5 (1):281–297.

Palsson, F., J. R. Sveinsson, and M. O. Ulfarsson. 2014. A new pansharpening algo-rithm based on total variation. *IEEE Geoscience and Remote Sensing Letters* 11 (1):318–322.

Palubinskas, G., P. Reinartz, and R. Bamler. 2010. Image acquisition geometry analy-sis for the fusion of optical and radar remote sensing data. *International Journal of Image and Data Fusion* 1 (3):271–282.

Park, J. H. and M. G. Kang. 2004. Spatially adaptive multi-resolution multispectral image fusion. *International Journal of Remote Sensing* 25 (23):5491–5508.

Pellemans, A., R. Jordans, and R. Allewijn. 1993. Merging multispectral and panchro-matic SPOT images with respect to the radiometric properties of the sensor. *Photogrammetric Engineering and Remote Sensing* 59 (1):81–87.

Pohl, C. 1996. Geometric aspects of multisensor image fusion for topographic map updating in the humid tropics. Dissertation, Institute of Photogrammetry, Leibnitz University Hannover, Hannover, Germany.

Pohl, C. and M. Hashim. 2014. Increasing the potential of Razaksat images for map-updating in the tropics. In *International Symposium on Digital Earth*, edited by M. Hashim. Kuching, Sarawak, Malaysia: IOP Conference Series: Earth and Environmental Science.

Pohl, C. and J. L. van Genderen. 1995. ERS-1 SAR and SPOT XS image fusion for topographic map updating in the tropics. Paper read at *2nd ERS Applications Workshop*, 6–8 December 1995, Queen Elizabeth Center, London, UK.

Pohl, C. and J. L. van Genderen. 1998. Review article. Multisensor image fusion in remote sensing: concepts, methods and applications. *International Journal of Remote Sensing* 19 (5):823–854.

Pohl, C. and J. L. van Genderen. 2015. Structuring contemporary remote sensing image fusion. *International Journal of Image and Data Fusion* 6 (1):3–21.

Pohl, C. and J. L. van Genderen. 1998. Review article. Multisensor image fusion in remote sensing: Concepts, methods and applications. *International Journal of Remote Sensing* 19 (5):823–854.

Pohl, C. and Y. Zeng. 2015. Development of a fusion approach selection tool. Paper read at *International Workshop on Image and Data Fusion*, 21–23 July 2015, Kona, Hawaii, USA.

Price, J. C. 1987. Combining panchromatic and multispectral imagery from dual reso-lution satellite instruments. *Remote Sensing of Environment* 21 (2):119–128.

Price, J. C. 1999. Combining multispectral data of differing spatial resolution. *IEEE Transactions on Geoscience and Remote Sensing* 37 (3):1199–1203.

Rahmani, S., M. Strait, D. Merkurjev, M. Moeller, and T. Wittman. 2010. An adap-tive IHS pan-sharpening method. *IEEE Geoscience and Remote Sensing Letters* 7 (4):746–750.

Ranchin, T., L. Wald, and M. Mangolini. 1996. The ARSIS method: A general solution for improving spatial resolution of images by the means of sensor fusion. In *Fusion of Earth Data*, edited by T. Ranchin and L. Wald, Cannes, France: EARSeL, pp. 53–58.

Rokach, L. 2016. Decision forest: Twenty years of research. *Information Fusion* 27:111–125.

Rong, K., S. Wang, S. Yang, and L. Jiao. 2014. Pansharpening by exploiting sharpness of the spatial structure. *International Journal of Remote Sensing* 35 (18):6662–6673.

Rothery, D. A. and P. W. Francis. 1987. Synergistic use of MOMS-01 and LANDSAT TM data. *International Journal of Remote Sensing* 8 (3):501–508.

Schowengerdt, R. A. 1980. Reconstruction of multispatial, multispectral image data using spatial frequency content. *Photogrammetric Engineering and Remote Sensing* 46 (10):1325–1334.

Schowengerdt, R. A. 2007a. Chapter 5—Spectral transforms. In *Remote Sensing*, 3rd edition, edited by R. A. Schowengerdt. Burlington: Academic Press.

Schowengerdt, R. A. 2007b. Chapter 6—Spatial transforms. In *Remote Sensing*, 3rd edition, edited by R. A. Schowengerdt. Burlington: Academic Press.

Schowengerdt, R. A. 2007c. Chapter 8—Image registration and fusion. In *Remote Sensing*, 3rd edition, edited by R. A. Schowengerdt. Burlington: Academic Press.

Selva, M., B. Aiazzi, F. Butera, L. Chiarantini, and S. Baronti. 2015. Hyper-sharpening: A first approach on SIM-GA Data. *IEEE Journal of Selected Topics in Applied Earth Observations and Remote Sensing* 8 (6):3008–3024.

Shah, V. P., N. H. Younan, and R. L. King. 2008. An efficient pan-sharpening method via a combined adaptive PCA approach and contourlets. *IEEE Transactions on Geoscience and Remote Sensing* 46 (5):1323–1335.

Sharpe, B., A. Kerr, and M. Dettwiler. 1991. Multichannel fusion technique for resolution enhancement during ground processing. Paper read at *Geoscience and Remote Sensing Symposium, 1991. IGARSS'91. Remote Sensing: Global Monitoring for Earth Management, International*, Espoo, Finland.

Siddiqui, Y. 2003. The modified IHS method for fusing satellite imagery. In *ASPRS Annual Conference*. Anchorage, Alaska.

Simone, G., A. Farina, F. C. Morabito, S. B. Serpico, and L. Bruzzone. 2002. Image fusion techniques for remote sensing applications. *Information Fusion* 3 (1):3–15.

Singh, A. 1989. Review Article. Digital change detection techniques using remotely-sensed data. *International Journal of Remote Sensing* 10 (6):989–1003.

Smith, C., M. Brown, R. Pouncey, K. Swanson, and K. Hart. 1999. *ERDAS Field Guide TM*. Atlanta, Georgia: ERDAS Inc.

Stefanski, J., O. Chaskovskyy, and B. Waske. 2014. Mapping and monitoring of land use changes in post-Soviet western Ukraine using remote sensing data. *Applied Geography* 55 (0):155–164.

Su, Y., C. H. Lee, and T. M. Tu. 2013. A multi-optional adjustable IHS-BT approach for high resolution optical and SAR image fusion. *Chung Cheng Ling Hsueh Pao/Journal of Chung Cheng Institute of Technology* 42 (1):119–128.

Tai, G. M. and S. I. Nipanikar. 2015. Implementation of image fusion techniques for remote sensing application. *International Journal of Emerging Technology and Advanced Engineering* 5 (6):109–113.

Te-Ming, T., P. S. Huang, H. Chung-Ling, and C. Chien-Ping. 2004. A fast intensity-hue-saturation fusion technique with spectral adjustment for IKONOS imagery. *IEEE Geoscience and Remote Sensing Letters* 1 (4):309–312.

Te-Ming, T., C. Wen-Chun, C. Chien-Ping, P. S. Huang, and C. Jyh-Chian. 2007. Best tradeoff for high-resolution image fusion to preserve spatial details and minimize color distortion. *IEEE Geoscience and Remote Sensing Letters* 4 (2): 302–306.

Tewes, A., F. Thonfeld, M. Schmidt et al. 2015. Using RapidEye and MODIS data fusion to monitor vegetation dynamics in semi-arid rangelands in South Africa. *Remote Sensing* 7 (6):6510–6534.

Thomas, C., T. Ranchin, L. Wald, and J. Chanussot. 2008. Synthesis of multispectral images to high spatial resolution: A critical review of fusion methods based on remote sensing physics. *IEEE Transactions on Geoscience and Remote Sensing* 46 (5):1301–1312.

Tu, T.-M., S.-C. Su, H.-C. Shyu, and P. S. Huang. 2001. A new look at IHS-like image fusion methods. *Information Fusion* 2 (3):177–186.

Vivone, G., L. Alparone, J. Chanussot et al. 2015. A critical comparison among pan-sharpening algorithms. *IEEE Transactions on Geoscience and Remote Sensing* 53 (5):2565–2586.

Vivone, G., R. Restaino, M. Dalla Mura, G. Licciardi, and J. Chanussot. 2014. Contrast and error-based fusion schemes for multispectral image pansharpening. *IEEE Geoscience and Remote Sensing Letters* 11 (5):930–934.

Wald, L. and T. Ranchin. 2003. The ARSIS concept in image fusion: An answer to users needs. Paper read at *Proceedings of the Sixth International Conference of Information Fusion*, Cairns, Queensland, Australia, July 8–11, 2003.

Xu, Q., Y. Zhang, and B. Li. 2014. Recent advances in pansharpening and key problems in applications. *International Journal of Image and Data Fusion* 5 (3): 175–195.

Xu, Q., Y. Zhang, B. Li, and L. Ding. 2015. Pansharpening using regression of classified MS and pan images to reduce color distortion. *IEEE Geoscience and Remote Sensing Letters* 12 (1):28–32.

Yang, X.-H. and L.-C. Jiao. 2008. Fusion algorithm for remote sensing images based on nonsubsampled contourlet transform. *Acta Automatica Sinica* 34 (3):274–281.

Yésou, H., Y. Besnus, and J. Rolet. 1993a. Extraction of spectral information from Landsat TM data and merger with SPOT panchromatic imagery—A contribution to the study of geological structures. *ISPRS Journal of Photogrammetry and Remote Sensing* 48 (5):23–36.

Yésou, H., Y. Besnus, J. Rolet, J. C. Pion, and A. Aing. 1993b. Merging Seasat and SPOT imagery for the study of geological structures in a temperate agricultural region. *Remote Sensing of Environment* 43 (3):265–279.

Zhang, C. 2015. Applying data fusion techniques for benthic habitat mapping and monitoring in a coral reef ecosystem. *ISPRS Journal of Photogrammetry and Remote Sensing* 104:213–223.

Zhang, J., J. Yang, Z. Zhao, H. Li, and Y. Zhang. 2010. Block-regression based fusion of optical and SAR imagery for feature enhancement. *International Journal of Remote Sensing* 31 (9):2325–2345.

Zhang, Y. and G. Hong. 2005. An IHS and wavelet integrated approach to improve pan-sharpening visual quality of natural colour IKONOS and QuickBird images. *Information Fusion* 6 (3):225–234.

Zhang, Y. and R. K. Mishra. 2014. From UNB PanSharp to Fuze Go—The success behind the pan-sharpening algorithm. *International Journal of Image and Data Fusion* 5 (1):39–53.

Zhou, X., J. Liu, S. Liu, L. Cao, Q. Zhou, and H. Huang. 2014. A GIHS-based spectral preservation fusion method for remote sensing images using edge restored spectral modulation. *ISPRS Journal of Photogrammetry and Remote Sensing* 88:16–27.

Zhou, Y. and F. Qiu. 2015. Fusion of high spatial resolution WorldView-2 imagery and LiDAR pseudo-waveform for object-based image analysis. *ISPRS Journal of Photogrammetry and Remote Sensing* 101:221–232.

Zhu, X., J. Chen, F. Gao, X. Chen, and J. G. Masek. 2010. An enhanced spatial and temporal adaptive reflectance fusion model for complex heterogeneous regions. *Remote Sensing of Environment* 114 (11):2610–2623.

Zobrist, A., R. Blackwell, and W. Stromberg. 1979. Integration of Landsat, Seasat, and other geo-data sources. Paper read at *13th International Symposium on Remote Sensing of Environment*, 23–27 April 1979, Ann Arbor, Michigan.

5

Quality Assessment

The entire effort of remote sensing and RSIF is to provide information about Earth and its complex processes. The information only makes sense if it is accompanied by a quality parameter allowing a proper judgment of its validity. Especially, if multiple sources and processing steps are involved, this quality assessment and error propagation is of essential importance. Successful RSIF aims at providing a fused image that inherits the properties and content of its input images. In RSIF, the assessment of the achievement and improvement of the process in the final fused product in terms of quality and contribution to the application is an active research field and the question whether or not an index represents fused image quality is still open (Zhang 2008). Over the years, researchers managed to establish a series of protocols for RSIF that contribute to an objective image quality evaluation. This chapter introduces the concept of quality and explains the various options of quality assessment, including the quantitative and qualitative approach. Of particular interest is the evaluation of pansharpening quality since it forms the most popular RSIF interest. For this purpose, a generic framework for pansharpening quality assessment is presented.

5.1 Overview

There is only a value in performing image fusion if the resulting fused image contains information of greater quality than the individual input images alone. This is reflected in the accepted definition of image fusion proposed by Pohl and Genderen (1998), and the work by Wald and his group contributed very valuable insights to the quality assessment aspect in image and data fusion (Wald et al. 1997; Thomas and Wald 2004). They refer to the issue of defining quality gain and using a genuine reference to obtain a quality statement. As a result, it is very important to distinguish between relative and absolute quality (Alparone et al. 2007; Thomas and Wald 2007). Relative quality refers to the closeness of the fused image to the original image using a verification of by defined ideal criteria, such as consistency and synthesis properties. Consistency refers to the possibility to reverse the pansharpening process by degrading the pansharpened image to obtain the original multispectral image. Synthesis describes the fact that the fused image should have

the characteristics of the original multispectral image at higher spatial resolution (Vivone et al. 2015b). Absolute quality corresponds to the similarity
between the fused and a reference image. The latter is produced synthetically
due to the fact that there is no access to the image that has the quality parameters for comparison. Different processes have established different quality
criteria. Classification usually relies on estimated Cohen's kappa statistic and
accuracy scores. Bias and goodness-of-fit statistics are used for regression
analysis. Image fusion per se uses quality indices providing information on
spatial and spectral integrity of the fused image (Gomez-Chova et al. 2015).

Until now, there is no established and generally accepted index that can
replace the human ability to assess the quality. The lack of an ideal reference image for comparison limits the objective evaluation of fusion results,
independent of the chosen assessment index (Vivone et al. 2015a). Therefore,
users of image fusion use two different ways to assess their achievements.
The first, a qualitative approach, is as the visible inspection of the results,
comparing the outcome to the original input data. The second, a quantitative approach, uses statistics and other assessment methods to provide comparable quality measures, such as quality indexes. An excellent overview
on general quality metrics is compiled in Li et al. (2010). Others measure
the outcome of an application of fused images, for example, in the context
of land cover mapping and classification (Colditz et al. 2006; Amarsaikhan
et al. 2011), which makes sense. Others suggest, both fusion performance
evaluation using quality measures and a classification accuracy assessment
if land cover maps are the anticipated output. They apply advanced classification algorithms to analyze spectral quality preservation in the fusion
process after having produced an *a priori* quality assessment with quality
indices (Yilmaz and Gungor 2015).

A visual evaluation of the results to assess fused image quality is performed in almost all published studies in the field. Visual interpretation
of the fusion results is necessary to identify local distortion of spectral and
spatial content of the image and to discover artifacts. We have dedicated a
separate section to visual interpretation following the description of quantitative measures (see Section 5.6). Visual analysis is important and in most of
the applications of RSIF the basis for the final decision. In military, defense,
and security applications, visual interpretation forms the major part of the
image analysis.

For the quantitative assessment, the research community has come up with
a variety of established quantitative evaluation criteria that can be applied
even though till today there is no unity on which criteria should be used as
standard (Thomas and Wald 2004; Zhang 2008). A selection of commonly
used indexes and assessment methods are listed in Table 5.1. This table is an
introduction to the subject and does not claim completeness. Furthermore,
indices and descriptions can be found in recent reviews and frameworks
on the subject (Li et al. 2010; Jagalingam and Hegde 2015; Palubinskas 2015;
Yilmaz and Gungor 2015). It illustrates the variety of quality measures image

TABLE 5.1

Quality Measures at a Glance

	Quality Index	Meaning of Abbreviation	Main Quality Assessed	Reference
With reference image PD	CC	Correlation coefficient	Spatial or spectral	Acerbi-Junior et al. (2006)
	$\Delta\mu$	Mean bias	Spectral	Han et al. (2008)
	PD	Per pixel deviation	Spectral and spatial	Wald (2002)
	RMSE	Root mean square error	General	Wald (2000)
	SAM	Spectral angle mapper	Spectral	Aiazzi et al. (2003)
	RASE	Relative average spectral error	Global spectral quality	Wald (2000)
	ERGAS	Erreur Relative Global Adimensionelle de Synthèse	Global spectral quality	Wald et al. (1997)
	UIQI[a]	Universal image quality index	Global spectral and spatial	Wang and Bovik (2002)
	(M)SSIM	(Mean) Structure similarity index measure	Luminance, contrast, and structure	Wang et al. (2004)
	SSQM	Structural similarity quality metric	Spatial	Wang et al. (2004)
	Q_{PS}	Quality pansharpening	Global spectral and spatial	Padwick et al. (2010)
Without reference image	QNR	Quality with no reference	Global spectral and spatial	Alparone et al. (2008)
	D	Degree of distortion	Global spectral and spatial	Zhu and Bamler (2013)
	σ	Standard deviation	Contrast	Karathanassi et al. (2007)
	H	Entropy	Richness of information	Han et al. (2008)

[a] Also called Q-average and also appears as Q in the literature.

fusion users have access to and the difficulty of choosing the "right" one. There is plenty of literature explaining the various indexes and applying them in image fusion studies (Gonzalez-Audicana et al. 2004; Wang et al. 2004; Alparone et al. 2007; Ling et al. 2007; Li et al. 2010; Meng et al. 2010; Padwick et al. 2010; Fonseca et al. 2011; Yilmaz and Gungor 2015). It should be noted that a lot of research on quality assessment reduces its focus on pansharpening only. The task of providing appropriate quantitative quality measurements is neither simple nor straightforward. That becomes obvious from the many publications in the field, showing that it is an active research field. Quality assessment is strongly influenced by the selection of quality indicators (Zhang 2008). It should be noted that there is no such thing as "the" quality index (Wang and Bovik 2009). Different measures or protocols deliver different statements, which need to be interpreted. The final decision even with the use of quantitative evaluation measures still remains with the operator and needs critical assessment and knowledge of the individual parameters, especially with respect to applications. For quantitative assessment, the presented protocols in Section 5.4 are state of the art, in particular the ones not requiring a reference image. For qualitative assessment, the description in Section 5.6 and Table 5.3 will assist to compare different RSIF results appropriately.

What we can also see from Table 5.1 is the fact that there is neither consensus in the literature on a uniform quality assessment approach, nor do we have a consistent naming for the various indexes. Looking at the mean structure similarity index measure (MSSIM) and the structural similarity quality metric (SSQM), for example, the different authors use similar approaches but the individual researchers keep producing their own terminology (Wang et al. 2004; Zheng and Qin 2009). Some signal quality measures do not allow a proper assessment. An adequate assessment is not possible with the MSE because it does not judge the image for the human perception of image fidelity (Wang and Bovik 2009).

Accepted protocols are a trustworthy and reliable source of quality statements. They pursue a combination of indices that produce a comparable parameter for quality. Therefore, we dedicate a separate section to these RSIF quality assessment protocols (see Section 5.3). These protocols have evolved because a single measure is not suitable to assess the quality of fused images. In particular, in the case of combining disparate data sets, for example, VIR and SAR data, the assessment has to consider the impact of the different information content of the imagery to reflect the effectiveness of a certain RSIF approach. All researchers agree that proper image quality assessment relies on both visual inspection and a quantitative approach.

5.2 Image Quality Parameters

Quantitative image quality assessment is based on mathematical modeling and often referred to as objective analysis (Jagalingam and Hegde 2015). Quality parameters or indicators to examine the spectral and spatial fidelity of the fused image are defined and then compared between input and output images. The goal is to determine the closeness of the two data sets or in other words their similarity. We distinguish two types, namely, with and without reference image as already shown in Table 5.1. In pansharpening, the reference image is the multispectral image at the resolution of the panchromatic image (Amro et al. 2011). The following sections explain the individual indices, indicating the quality parameter they examine and the necessary information needed for evaluation.

5.3 Acceptable Quality Measures in RSIF

The list of indices to quantitatively assess fused remote sensing images is long. In this section, we explain the most common metrics, including the definition of statistical measures that are used in more complex indices and protocols. The individual measure (e.g., MSE, RMSE, CC) might not deliver an appropriate quality statement but contributes to the global assessment using the complex indices (e.g., RASE, ERGAS). The equations contain variables that are explained in Table 5.2.

5.3.1 Mean Bias

Mean bias ($\Delta\mu$) is a measure of the shift of the mean value of the histogram due to the fusion process and is popular in pansharpening quality assessment. It is an indicator of changes to the histogram of the fused image compared to the original multispectral image. It shows a shift to white if its value is positive. In case of a negative bias of the mean value of the histogram of the fused image, there is a shift to gray (Jawak and Luis 2013). The ideal value of $\Delta\mu$ is zero. It is calculated as the difference between the two means as shown in Equation 5.1:

$$\Delta\mu = \mu(f) - \mu(r) \tag{5.1}$$

TABLE 5.2

Variable Definitions in Equations for Quality Indices

Variable	Definition
f	Fused image
r	Reference image
m	Number of image lines
n	Number of image columns
x, y	Pixel location in the image
σ	Standard deviation
$\sigma_{f,r}$	Covariance of fused and reference image
b	Number of multispectral bands
d_h	Pixel size high-resolution image
d_l	Pixel size low-resolution image

5.3.2 Mean Square Error

Spectral distortion between the fused and the reference (original) image is measured as MSE. Its ideal value is zero, which means $f = r$. Its mathematical expression is listed in Equation 5.2:

$$\text{MSE} = \frac{\sum_{x=1}^{m}\sum_{y=1}^{n}(f(x,y)-r(x,y))^2}{m*n} \tag{5.2}$$

MSE is not suitable to estimate image quality in terms of human visual perception. Distorted image containing obviously rather different errors that can be recognized by visual inspection can result in a nearly identical MSE value. In order to judge image quality for visual image analysis, the structure similarity index metric (SSIM) discussed in Section 5.3.11 is a more suitable approach (Wang and Bovik 2009).

5.3.3 Root Mean Square Error

Root mean square error (RMSE) provides the standard error of fused images and expresses the spatial and spectral distortion contained in the fused image. It is very popular even though the individual band errors do not relate to the band's mean value. RMSE often match visual evaluation results and vegetation indices (Vivone et al. 2015a). It is expressed in Equation 5.3:

$$\text{RMSE} = \sqrt{\text{MSE}} \tag{5.3}$$

An optimum RMSE is zero, meaning that $f = r$.

5.3.4 Normalized Root Mean Square Error

The normalized root mean square error (NRMSE) also represents distortion spatial and spectral distortion. Equation 5.4 is

$$\text{NRMSE} = \frac{\sum_{x=1}^{m}\sum_{y=1}^{n}(f(x,y)-r(x,y))^2}{\sum_{x=1}^{m}\sum_{y=1}^{n}(r(x,y))^2} \tag{5.4}$$

5.3.5 Correlation Coefficient

Correlation coefficient (CC) provides the degree of correlation between the fused and the reference image. It is also called cross-correlation coefficient and belongs to the group of similarity indices. Ranging between [–1,1], its best value is one; meaning that the compared images are highly correlated, that is, coincide and only differ by their global mean offset and gain factor. A value of –1 means that one image is the inverted version of the other (Aiazzi et al. 2012). Equation 5.5 describes how the CC is calculated:

$$CC = \frac{\sum_{x=1}^{m}\sum_{y=1}^{n}(f(x,y)-\mu_f)(r(x,y)-\mu_r)^2}{\sqrt{\sum_{x=1}^{m}\sum_{y=1}^{n}((f(x,y)-\mu_f)^2(r(x,y)-\mu_r)^2)}} \tag{5.5}$$

5.3.6 Universal Image Quality Index

As similarity index identifying spectral and spatial distortions, it uses covariance, variances, and means of fused and reference images in Equation 5.6:

$$UIQI = \frac{\sigma_{fr}}{\sigma_f\sigma_r} * \frac{2\mu_f\mu_r}{\mu_f^2+\mu_r^2} * \frac{2\sigma_f\sigma_r}{\sigma_f^2+\sigma_r^2} \tag{5.6}$$

Therefore, it is a measure that reflects correlation loss (first component = CC, range [–1.1]), distortions in luminance (second component, range [0,1], best value 1), and contrast distortions (third component, range [0,1], best value 1) (Wang and Bovik 2002).

5.3.7 UIQI for Four Band Imagery

An adaptation of the UIQI for multispectral images of four bands, for example, QuickBird or IKONOS multispectral bands is expressed by four band imagery (Q4). It considers local mean bias, contrast variations, loss of correlation, and spectral distortions and is based on the quaternion theory, which can only consider four bands as it uses hypercomplex correlation (Alparone et al. 2004). Its range reaches [0,1], while its best value is one, meaning that the fused and the reference image are identical. It is possible to provide local quality expressions if Q_4 is calculated for $k \times k$ windows with Equation 5.7 and can then be averaged to one global Q_4-value. k refers to the kernel size. Summarizing the mathematical operations, the equation can also be expressed as Equation 5.8.

$$Q_4 = \frac{|\sigma_{fr}|}{\sigma_f\sigma_r} * \frac{2|\mu_f|*|\mu_r|}{|\mu_f|^2+|\mu_r|^2} * \frac{2\sigma_f\sigma_r}{\sigma_f^2+\sigma_r^2} \tag{5.7}$$

$$Q_4 = \frac{4*|\sigma_{fr}|*|\mu_f|*|\mu_r|}{(\sigma_f^2+\sigma_r^2)*(|\mu_f|^2+|\mu_r|^2)} \tag{5.8}$$

5.3.8 Spectral Angle Mapper

SAM expresses the spectral similarity between two spectra (two images) by calculating the angle between the images. The principle is depicted in Figure 5.1. These are considered vectors with b dimensions, b representing the number of bands.

This technique, when used on calibrated reflectance data, is relatively insensitive to illumination and albedo effects. As spectral vector measure SAM provides the angle between corresponding pixels of the fused and reference images. The compared spectral bands form the coordinate axes (Vivone et al. 2015a). Its value is expressed in degrees or radians with its best achievable value being zero. It is calculated using Equation 5.9. It is the absolute value between two spectral vectors and averaged to yield a global measure for the entire image (Bovolo et al. 2010). It is interesting to know that even in the absence of spectral distortion (SAM = 0), radiometric distortions still exist. Practically, the two vectors have different lengths but are parallel (Nencini et al. 2007).

$$\text{SAM}(f,r) = \arccos\left(\frac{\langle f,r\rangle}{\|f\|_2 * \|r\|_2}\right) \tag{5.9}$$

5.3.9 Relative Average Spectral Error

Relative average spectral error (RASE) is another global index using the average M of the mean per band i for b number of bands and is expressed in percent resulting from Equation 5.10. This index provides a quick overview in

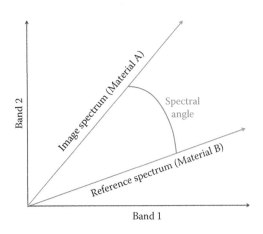

FIGURE 5.1
SAM concept.

RSIF algorithm comparison, where a lower value means better image quality. RASE is provided in percent, and its ideal value is zero.

$$\text{RASE} = \frac{100}{\mu} \sqrt{\frac{1}{b} \sum_{i=1}^{b} \text{RMSE}(f_i, r_{i_i})} \qquad (5.10)$$

5.3.10 Relative Dimensionless Global Error of Synthesis

ERGAS* is a global index and sums up RMSE values of the individual bands (Wald 2000) as indicated in Equation 5.11. Its ideal value is zero because it forms the sum of the RMSE values (Vivone et al. 2015a), and it is expressed in percent. It considers the scale relationship (ratio) between the two images to be fused.

$$\text{ERGAS} = 100 \frac{d_h}{d_l} \sqrt{\frac{1}{b} \sum_{i=1}^{b} \frac{\text{RMSE}(f_i, r_i)}{\mu(f_i)}} \qquad (5.11)$$

5.3.11 Structure Similarity Index Metric

As the name already states, it estimates the structural similarity between the fused and the reference image (Wang et al. 2004). The algorithm is shown in Equation 5.12. The constants C_1 and C_2 are necessary to avoid a division by zero. They depend on the dynamic range of the pixel values.

$$\text{SSIM} = \frac{(2\mu_f \mu_r + C_1) * (2\sigma_f \sigma_r + C_2)}{(\mu_f^2 + \mu_r^2 + C_1) * (\sigma_f^2 + \sigma_r^2 + C_2)} \qquad (5.12)$$

SSIM compares luminance, contrast, and structure using means and standard deviations of fused and reference image. The idea to develop this measure came from the assumption that the human vision prioritizes structural information extraction (Wang et al. 2004). The higher the value of the measure, the better is the expected quality of the fused image.

5.4 RSIF Quality Assessment Protocols

The idea of establishing quality assessment protocols in RSIF arose from the need for criteria to quantitatively evaluate spectral and spatial integrity of fused images. Researchers found that individual indices do not reflect the

---

* Relative dimensionless global error of synthesis: the abbreviation is derived from the French term "Erreur Relative Globale Adimensionnelle de Synthèse."

quality of fused images sufficiently since they only relate to a single criterion or are not sensitive to aspects that are relevant in the assessment of fused multimodal imagery.

5.4.1 Wald's Protocol

The conditions for an appropriate quantitative assessment were defined for pansharpening in 1997, containing three properties (Wald et al. 1997). The fused (pansharpened) image should be as similar as possible to

1. The original multispectral image if resampled to the lower multi-spectral image resolution (consistency); this measures spectral quality after spatial enhancement
2. The ideal image that a corresponding sensor would observe at the same high spatial resolution (synthesis—scalar/per band)
3. The multispectral part (vector) of the multispectral part (vector) of an ideal image acquired by a corresponding sensor at the same high spatial resolution (synthesis—vector/all bands)

The synthesis condition is therefore split into two properties (second and third). Both are difficult to implement since usually such an ideal image does not exist. Therefore, synthesis is analyzed using downscaled images. The spatial degradation is reached on the basis of low-pass filtering applying a factor based on the scale ratio of the two input images (Aiazzi et al. 2012). The actual fusion process is carried out on the modified images, which leads to fused images at the lower spatial resolution. These can then be compared to the original lower-resolution multispectral image, which represents the "ideal" reference image for conditions two and three.

5.4.2 Zhou's Protocol

This protocol goes back to the 1990s and separates spectral and spatial quality. Spectral quality is based on the assumption that the spectral content of the fused image should be similar to the spectral content of the original multispectral image. It refers to the fact that spectral signatures of certain targets are maintained. Practically, the metric computes the average differences between pixel values of the fused (f) and original, resampled multispectral image (MS), band by band following Equation 5.13:

$$D_i = \frac{1}{m*n}\sum_R\sum_C |f_{iRC} - \mathrm{MS}_{iRC}| \qquad (5.13)$$

valid for the Rth row and Cth column of the ith band. In an ideal case, the difference is zero. The spatial quality is quantitatively evaluated on the assumption that spatial information is unique in the high-resolution pan-chromatic image in the high frequencies. Calculating the correlation between the high-pass filtered versions of the fused and panchromatic images tests the spatial quality of the fused image. The filtering is performed on the basis of a Laplacian filter (Zhou et al. 1998). Another common name of this protocol is spatial correlation index (SCI) or high-pass correlation coefficient (HCC), and its ideal value should be one (highly correlated). Drawbacks of Zhou's protocol are the fact that the results in contradictory quality statements in the spatial domain if compared to quality metrics with reference images, such as Q4, ERGAS, and SAM (Alparone et al. 2008).

5.4.3 Quality with No Reference Protocol

The lack of a reference image for quantitative quality assessment has led to the development of another evaluation method using the QNR protocol a decade later. One constraint is the unavailability of a reference image, and the other, the assumption that fusion algorithms operate scale invariant if evaluation is done at coarser scales. This might not be true for very high-resolution imagery of complex environments. The global quality assessment protocol QNR works at full scale and its two parts separately assess spatial and spectral quality of the fused image (Alparone et al. 2008). The basis for QNR spectral and spatial quality is the UIQI similarity assessment of scalar image pairs. The idea is to compare the fused image bands with the origi-nal multispectral image (spectral quality) and with the panchromatic band (spatial quality). Spectral distortions are determined between the fused and the original multispectral image using Equation 5.14, at both resolutions: (1) original resolution of multispectral image (MS, PAN$_{low}$) and (2) high reso-lution of the original panchromatic band (PAN) and the fused (pansharp-ened) image (f). The spatial component is produced by Equation 5.15, again at both resolutions. Owing to the interband differences at two scales, it is possible to analyze distortion across scales (Khan et al. 2008).

$$D_{spectral} = \sqrt[p]{\frac{1}{b(b-1)} \sum_{i=1}^{b} \sum_{k=1, k\neq i}^{b} UIQI(MS_i, MS_k) - UIQI(f_i, f_k)} \qquad (5.14)$$

$$D_{spatial} = \frac{1}{b} \sum_{i=1}^{b} |UIQI(MS_i, PAN_{low}) - UIQI(f_i, PAN)| \qquad (5.15)$$

with MS$_i$ representing the ith original multispectral band and MS$_k$ one of the other original multispectral bands to be compared with. Similarly, f_i is the ith, and f_k is the kth band of the fused image. A tuning parameter p allows the weighting of spectral differences, which means, for $p = 1$, all differences are given equal weight, and for $p > 1$, large spectral differences are weighted higher. The resulting ideal value of $D_{spectral}$ and $D_{spatial}$ is zero.

The joint quality measure QNR is then derived from Equation 5.16:

$$QNR = (1 - D_{spectral})^{\alpha} - (1 - D_{spatial})^{\beta} \tag{5.16}$$

The combined global measure should therefore reach 1 in the best case. QNR can be applied locally using image blocks to obtain a more detailed quantitative analysis (Khan et al. 2008). α and β are further tuning parameters to focus the quality assessment on either spectral or spatial components of the fused image. In the case of $\alpha = \beta = 1$, both aspects are equally considered in the evaluation process.

5.4.4 Khan's Protocol

This protocol evolved by taking advantage of individual aspects of the three protocols explained above. It combines the consistency property (Wald's protocol), the spatial component of Zhou's protocol, and the spectral distortion from QNR. Khan's protocol produces two indices, that is, spectral (consistency) and spatial quality (modified HCC). Instead of using Laplacian filters for the low- and high-frequency information extraction this protocol uses filters adapted to the MTF of the individual bands (Khan et al. 2009).

This protocol assesses spectral and spatial quality of the fused image separately. The spectral index uses an MTF filter to reduce the resolution of the pansharpened image to the original multispectral image resolution and compares the resulting degraded image with the original image, using Q4. The resulting difference forms the spectral distortion index (see Figure 5.2).

For spatial quality assessment, this protocol uses MTF filters for high-frequency information extraction from the original and fused resolutions plus the PAN in its downscaled version. UIQI delivers the measure for the comparison of spatial details from MS and the details of PAN at both resolutions. The flow of Khan's protocol to calculate the spatial distortion index is illustrated in Figure 5.3.

In practical applications, it is suggested to combine Zhou's protocol with QNR to obtain two objective evaluation methods for a final decision (Khan et al. 2009).

5.5 Palubinskas' Generic Framework for Pansharpening Quality Assessment

Palubinskas (2014) found out that MSE- and UIQI/SSIM-based measures are unsuitable to assess image quality with respect to visual image quality

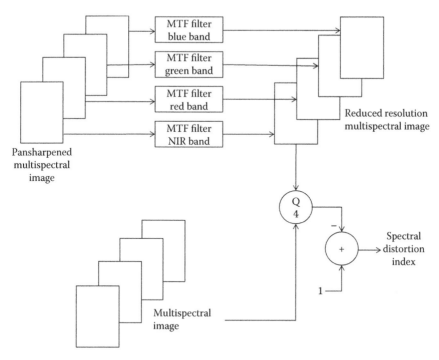

FIGURE 5.2
Khan's protocol to produce the spectral distortion index. (Adapted from Khan, M. M. et al. 2009. *IEEE Transactions on Geoscience and Remote Sensing*, 47(11): 3880–3891.)

perception. These indices are not suitable for translation invariance with respect to sample means and standard deviation applications. He proposed a new measure, forming a composite based on means, standard deviations, and correlation coefficients (CMSC). This measure is translation invariant with respect to means and standard deviations. This section introduces this new measure and discusses its benefits.

Two important properties of adequate distance (d) measures are translation and scale invariance (Palubinskas 2015). Translation invariance is expressed in Equation 5.17:

$$d(p_1 + c, p_2 + c) = d(p_1, p_2) \tag{5.17}$$

Scale invariance is defined by Equation 5.18:

$$d(c * p_1, c * p_2) = d(p_1, p_2) \tag{5.18}$$

p_i are variables; c forms a constant.

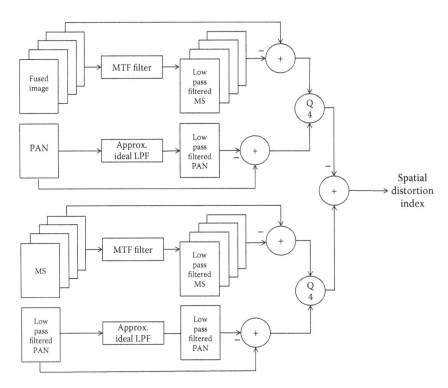

FIGURE 5.3
Flowchart to produce the spatial distortion index according to Khan's protocol. (Adapted from Khan, M. M. et al. 2009. *IEEE Transactions on Geoscience and Remote Sensing*, 47(11): 3880–3891.)

Equation 5.17 leads to Equation 5.19:

$$d(p, p+c) = \text{constant} \tag{5.19}$$

for all p and a fixed c.

CC is a translation- and scale-invariant measure. MSE is translation invariant. UIQI or SSIM are scale invariant but not translation invariant, which can lead to problems in case of classification, clustering, matching, and change detection (Palubinskas 2015). Instead, a translation-invariant measure based on means and standard deviations can be used, formulated in Equations 5.20 through 5.22:

$$\text{CMSC}(f, r) = (1 - d_1) * (1 - d_2) * \rho \tag{5.20}$$

$$d_1 = \frac{(\mu_f \mu_r)^2}{R^2} \tag{5.21}$$

$$d_2 = \frac{(\sigma_f - \sigma_r)^2}{(R/2)^2} \tag{5.22}$$

where ρ is Pearson's correlation coefficient and $R = 2^8 - 1 = 255$ for 8 bit data. The optimum value for CMSC is one.

5.6 Value and Requirements for Visual Evaluation

Until today, visual interpretation of satellite images is relevant. There is no automated procedure or technology that can replace human vision. Geological applications rely heavily on visual interpretation of fused multi-sensor imagery and form only one example where the automated processes are part of the preprocessing (see also Chapter 6). Even though visual quality assessment has a subjective component, it still remains extremely important. If the purpose of image fusion is visual interpretation, for example, to take a decision based on the data like in security-relevant actions, a visual assessment is sufficient. It is indispensible in other applications too because a solely automated quantitative assessment is not reliable. There is no quality index that can replace human vision and perception.

Visual quality assessment has to take place under similar visualization conditions, which include stretching and display (Zhang 2008) if the result should be comparable and repeatable. RSIF quality criteria by visual interpretation completely differ from quantitative measures because it is neither always necessary nor useful to produce images with the same spectral or spatial characteristics of the input data. This is in particularly valid in case of VIR/SAR fusion where the radiometric values could change drastically but an interesting texture and 3D optical component can be introduced, which significantly increase interpretation capability. Two examples for such imagery are provided in Chapter 6 where applications of RSIF are discussed (cf. Figures 6.9 and 6.15).

Relevant elements for image interpretation are scale (size of objects, image resolution), patterns, contrast, color, tones, and shadows. In terms of color, the IHS representation of color is closed to human understanding (Malpica 2007). Multi-temporal images require an understanding of monitored processes so that differences can be assessed appropriately in the interpretation process. Preprocessing and RSIF can influence all these features. A major factor is the experience of the image analyst, also depending on the type of sensor. New types of sensors introduce new challenges in image interpretation. The visual interpretation of fused images requires knowledge about all sensors and data involved in the fusion process, including the fusion technique, the scene, and the application. Relevant reference data play a key role, especially existing maps and *in situ* data. If these are available, an image quality assessment is reliable. Visual interpretation also benefits strongly of the third dimension. If image data can provide stereo viewing capabilities or the data can be represented as 3D model with an underlying DEM, image interpretation is facilitated.

For visual interpretation, the resampling algorithm is of importance. Even though cubic convolution might result in a larger difference to the original image acquisition, there are fewer disturbances by artificial structures (set-back) induced by the resampling process like with nearest neighborhood resampled images (see Figure 5.4).

The image analyst can assign grades of image quality using absolute and relative measures in visual interpretation. Table 5.3 shows the measures

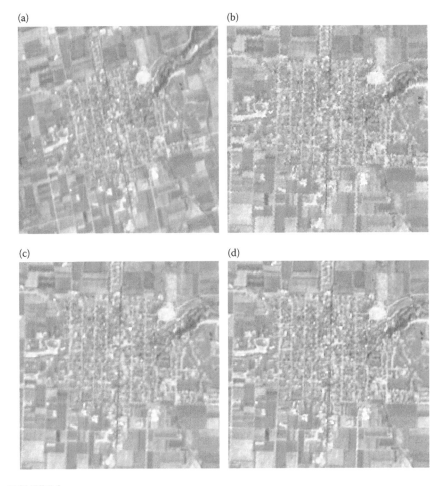

FIGURE 5.4
Resampling effect: (a) original SPOT XS image; (b) nearest neighbor; (c) bilinear; (d) cubic convolution. The images are resampled for geocoding.

TABLE 5.3

Qualitative Image Quality Evaluation

Grade	Measure (Absolute)	Measure (Relative)
1	Excellent	Best
2	Good	Better than average
3	Fair	Average
4	Poor	Lower than average
5	Very poor	Lowest

Source: Shi, W. et al. 2005. *International Journal of Applied Earth Observation and Geoinformation,* 6(3–4): 241–251.

introduced by Shi et al. (2005). Sometimes, quality measures for visual interpretations are designed to suit a certain organization or application to make fusion results in an operational environment compatible (Pohl and Touron 2000).

5.7 Summary

As already stated, the selection of quality indicators is not as straightforward as one would think. In some cases, researchers disagree on the validity of indices. Even though ERGAS has been suggested as global image quality index with the ability to provide spectral quality measures, other authors state that ERGAS does not provide spectral quality and therefore Q4 should be used (Alparone et al. 2004). The great disadvantage of Q4, however, is its limitation to four spectral bands. The advantage of ERGAS is its independence of units, spectral bands, and ratio of scales (Wald 2000). If we consider the discussion of Zhang (2008), apparently no convincing global quality measure exists. Therefore, a quality assessment requires a proper definition of quality parameters and corresponding measures that succeed to represent the quality of these parameters. This might include a series of measures. In conclusion, we can state that there is no such thing as the one for all image quality measure and visual assessment under comparable conditions in terms of display conditions and stretching is still necessary.

References

Acerbi-Junior, F. W., J. G. P. W. Clevers, and M. E. Schaepman. 2006. The assessment of multi-sensor image fusion using wavelet transforms for mapping the Brazilian Savanna. *International Journal of Applied Earth Observation and Geoinformation* 8 (4):278–288.

Aiazzi, B., L. Alparone, S. Baronti, A. Garzelli, and M. Selva. 2003. An MTF-based spectral distortion minimizing model for pan-sharpening of very high resolution multispectral images of urban areas. Paper read at *Remote Sensing and Data Fusion over Urban Areas, 2003. 2nd GRSS/ISPRS Joint Workshop on*, 22–23 May 2003, Berlin, Germany.

Aiazzi, B., L. Alparone, S. Baronti, A. Garzelli, and M. Selva. 2012. Twenty-five years of pansharpening: A critical review and new developments. In *Signal and Image Processing for Remote Sensing*, edited by C. H. Chen. Boca Raton, FL: CRC Press.

Alparone, L., B. Aiazzi, S. Baronti, A. Garzelli, F. Nencini, and M. Selva. 2008. Multispectral and panchromatic data fusion assessment without reference. *Photogrammetric Engineering & Remote Sensing* 74 (2):193–200.

Alparone, L., S. Baronti, A. Garzelli, and F. Nencini. 2004. A global quality measurement of pan-sharpened multispectral imagery. *IEEE Transactions on Geoscience and Remote Sensing* 1 (4):313–317.

Alparone, L., L. Wald, J. Chanussot, C. Thomas, P. Gamba, and L. M. Bruce. 2007. Comparison of pansharpening algorithms: Outcome of the 2006 GRS-S data-fusion contest. *IEEE Transactions on Geoscience and Remote Sensing* 45 (10):3012–3021.

Amarsaikhan, D., M. Saandar, M. Ganzorig et al. 2011. Comparison of multisource image fusion methods and land cover classification. *International Journal of Remote Sensing* 33 (8):2532–2550.

Amro, I., J. Mateos, M. Vega, R. Molina, and A. Katsaggelos. 2011. A survey of classical methods and new trends in pansharpening of multispectral images. *EURASIP Journal on Advances in Signal Processing* 2011 (1):79.

Bovolo, F., L. Bruzzone, L. Capobianco, A. Garzelli, S. Marchesi, and F. Nencini. 2010. Analysis of the effects of pansharpening in change detection on VHR images. *IEEE Geoscience and Remote Sensing Letters* 7 (1):53–57.

Colditz, R. R., T. Wehrmann, M. Bachmann et al. 2006. Influence of image fusion approaches on classification accuracy: A case study. *International Journal of Remote Sensing* 27 (15):3311–3335.

Fonseca, L., L. Namikawa, E. Castejon, L. Carvalho, C. Pinho, and A. Pagamisse. 2011. Image fusion for remote sensing applications. In *Image Fusion and Its Applications*, edited by Y. Zheng. Rijeka, Croatia: InTech.

Gomez-Chova, L., D. Tuia, G. Moser, and G. Camps-Valls. 2015. Multimodal classification of remote sensing images: A review and future directions. *Proceedings of the IEEE* 103 (9):1560–1584.

Gonzalez-Audicana, M., J. L. Saleta, R. G. Catalan, and R. Garcia. 2004. Fusion of multispectral and panchromatic images using improved IHS and PCA mergers based on wavelet decomposition. *IEEE Transactions on Geoscience and Remote Sensing* 42 (6):1291–1299.

Han, S., H. Li, and H. Gu. 2008. The study on image fusion for high spatial resolution remote sensing images. *The International Archives of the Photogrammetry, Remote Sensing and Spatial Information Sciences* 37:1159–1163.

Jagalingam, P. and A. V. Hegde. 2015. A review of quality metrics for fused image. In *International Conference on Water Resources, Coastal and Ocean Engineering*, edited by G. S. Dwarakish. Amsterdam, Netherlands: Elsevier.

Jawak, S. D. and A. J. Luis. 2013. A comprehensive evaluation of PAN-sharpening algorithms coupled with resampling methods for image synthesis of very high resolution remotely sensed satellite data. *Advances in Remote Sensing* 2:332–344.

Karathanassi, V., P. Kolokousis, and S. Ioannidou. 2007. A comparison study on fusion methods using evaluation indicators. *International Journal of Remote Sensing* 28 (10):2309–2341.

Khan, M. M., L. Alparone, and J. Chanussot. 2009. Pansharpening quality assessment using the modulation transfer functions of instruments. *IEEE Transactions on Geoscience and Remote Sensing* 47 (11):3880–3891.

Khan, M. M., J. Chanussot, B. Siouar, and J. Osman. 2008. Using QNR index as decision criteria for improving fusion quality. Paper read at *2nd International Conference on Advances in Space Technologies, 2008. ICAST 2008*, Islamabad, Pakistan.

Li, S., Z. Li, and J. Gong. 2010. Multivariate statistical analysis of measures for assessing the quality of image fusion. *International Journal of Image and Data Fusion* 1 (1):47–66.

Ling, Y., M. Ehlers, E. L. Usery, and M. Madden. 2007. FFT-enhanced IHS transform method for fusing high-resolution satellite images. *ISPRS Journal of Photogrammetry and Remote Sensing* 61 (6):381–392.

Malpica, J. A. 2007. Hue adjustment to IHS pan-sharpened IKONOS imagery for vegetation enhancement. *IEEE Geoscience and Remote Sensing Letters* 4 (1):27–31.

Meng, Q., B. Borders, and M. Madden. 2010. High-resolution satellite image fusion using regression kriging. *International Journal of Remote Sensing* 31 (7):1857–1876.

Nencini, F., A. Garzelli, S. Baronti, and L. Alparone. 2007. Remote sensing image fusion using the curvelet transform. *Information Fusion* 8 (2):143–156.

Padwick, C., M. Deskevich, F. Pacifici, and S. Smallwood. 2010. Worldview-2 pansharpening. Paper read at *ASPRS Annual Conference*, San Diego, California.

Palubinskas, G. 2014. Mystery behind similarity measures MSE and SSIM. Paper read at *2014 IEEE International Conference on Image Processing (ICIP)*, Paris, France, October 27–30.

Palubinskas, G. 2015. Joint quality measure for evaluation of pansharpening accuracy. *Remote Sensing* 7 (7):2072–4292.

Pohl, C. and H. Touron. 2000. Issues and challenges of operational applications using multisensor image fusion. In *3rd International Conference on Fusion of Earth Data on Fusion of Earth Data: Merging Point Measurements, Raster Maps and Remotely Sensed Images*, edited by T. Ranchin and L. Wald. Sophia Antipolis, France: SEE/URISCA, pp. 25–31.

Pohl, C. and J. L. van Genderen. 1998. Review article multisensor image fusion in remote sensing: Concepts, methods and applications. *International Journal of Remote Sensing* 19 (5):823–854.

Shi, W., C. Zhu, Y. Tian, and J. Nichol. 2005. Wavelet-based image fusion and quality assessment. *International Journal of Applied Earth Observation and Geoinformation* 6 (3–4):241–251.

Thomas, C. and L. Wald. 2004. Assessment of the quality of fused products. Paper read at *New Strategies for European Remote Sensing. Proceedings of the 24th Symposium of the European Association of Remote Sensing Laboratories (EARSeL)*, 25–27 May 2004, Dubrovnik, Croatia.

Thomas, C. and L. Wald. 2007. Comparing distances for quality assessment of fused images. Paper read at *26th EARSeL Symposium New Developments and Challenges in Remote Sensing*, 29 May–2 June 2006, Varsovie, Poland.

Vivone, G., L. Alparone, J. Chanussot et al. 2015a. A critical comparison among pansharpening algorithms. *IEEE Transactions on Geoscience and Remote Sensing* 53 (5):2565–2586.

Vivone, G., M. Simoes, M. Dalla Mura et al. 2015b. Pansharpening based on semi-blind deconvolution. *IEEE Transactions on Geoscience and Remote Sensing* 53 (4):1997–2010.

Wald, L. 2000. Quality of high resolution synthesised images: Is there a simple criterion? Paper read at *Fusion of Earth Data: Merging Point Measurements, Raster Maps and Remotely Sensed Images*, 26–28 January 2000, Sophia Antipolis, France.

Wald, L. 2002. Data fusion. Definitions and architectures. Fusion of images of different spatial resolutions. http://www.ebookdb.org/reading/2FGD1E4422461C3 83E5F7F69/Data-Fusion—Definitions-And-Architectures—Fusion-Of-Images-Of-Different-Spatia: http://www.ensmp.fr/Presses.

Wald, L., T. Ranchin, and M. Mangolini. 1997. Fusion of satellite images of different spatial resolutions—Assessing the quality of resulting images. *Photogrammetric Engineering & Remote Sensing* 63 (6):691–699.

Wang, Z. and A. C. Bovik. 2002. A universal image quality index. *IEEE Signal Processing Letters* 9 (3):81–84.

Wang, Z. and A. C. Bovik. 2009. Mean squared error: Love it or leave it? A new look at signal fidelity measures. *IEEE Signal Processing Magazine* 26 (1):98–117.

Wang, Z., A. C. Bovik, H. R. Sheikh, and E. P. Simoncelli. 2004. Image quality assessment: From error visibility to structural similarity. *IEEE Transactions on Image Processing* 13 (4):600–612.

Yilmaz, V. and O. Gungor. 2015. Fusion of very high-resolution UAV images with criteria-based image fusion algorithm. *Arabian Journal of Geosciences* 9 (1):1–16.

Zhang, Y. 2008. Methods for image fusion quality assessment—A review, comparison and analysis. Paper read at 37th ISPRS Congress, Beijing, China: ISPRS, pp. 1101–1109.

Zheng, Y. and Z. Qin. 2009. Objective image fusion quality evaluation using structural similarity. *Tsinghua Science & Technology* 14 (6):703–709.

Zhou, J., D. L. Civco, and J. A. Silander. 1998. A wavelet transform method to merge Landsat TM and SPOT panchromatic data. *International Journal of Remote Sensing* 19 (4):743–757.

Zhu, X. X. and R. Bamler. 2013. A sparse image fusion algorithm with application to pan-sharpening. *IEEE Transactions on Geoscience and Remote Sensing* 51 (5):2827–2836.

6

Applications

This important chapter contains practical experiences of remote sensing image fusion. It aims at providing an overview on existing practices. This overview on implemented applications for remote sensing image fusion was compiled from published results from international peer-reviewed journals and complemented by expert experiences. For the past 23 years, the authors themselves carried out quite a number of studies in RSIF, among which are applications such as topographic map updating (Pohl 1996; Pohl and Genderen 1999; Pohl and Hashim 2014), mineral exploitation (Pour et al. 2013), underground coal fires (Zhang et al. 1998), urban mapping (Amarsaikhan et al. 2010), landmine/minefield detection (Maathuis and Genderen 2004a,b), land use/land cover mapping (Zeng et al. 2010), and earthquake prediction (Genderen 2004, 2005). The identification of achievements in RSIF for different applications is rather difficult because more research has gone into the development of better algorithms rather than advancing the application. The transfer from science to application and operational implementation still lacks attention even though quite some advancement has been achieved in the past decade. This chapter intends to draw attention to this fact and provides a good overview on the current state of the art in RSIF applications. Most applications can be observed at local, regional, and global scales. The choice of area coverage strongly influences all other parameters such as sensor selection, data preprocessing, and fusion technique followed by postprocessing and data exploitation.

It is rather interesting to learn that the topics concentrate on global problems that reflect the current challenges that humanity faces. One quarter of the applications reported on belong to the study of urbanization processes. Agriculture and change detection are other important applications as shown in Figure 6.1 (Pohl 2016).

Both applications reflect that the topic is of global interest. The papers mainly focus on anthropogenic aspects, such as the development of megacities and the change in land use, that is, conversion of natural landscapes into industrial and agricultural areas. Remote sensing image fusion is a major contributor in image processing due to the increased spatial resolution of available sensors in combination with pansharpening, which makes urban mapping from space feasible. Both applications are key players in the studies about climate change. The distribution of topics among the subsections might have some overlap, such as land cover/land use mapping and urban studies or vegetation and forestry. However, it was considered necessary to

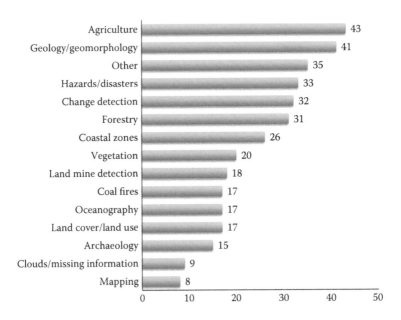

FIGURE 6.1
Applications of RSIF as published in international, peer-reviewed journals.

split the information into more detail to focus on the particular parameters required for the application. If there is a strong overlap in the fusion methods used, the information is linked up with the other subsection in question.

RSIF processing and techniques are described below in relation to the various applications, among them are urbanization, land use/land cover, change detection, geology, vegetation, agriculture, forest, hazards/disasters, oceanography, and missing information. In general, there are two trends one can observe when looking at current practices. The first one relates to the increase in spatial resolution prior to the final analysis. Owing to the availability of very high spatial resolution images, pansharpening is a popular prerequisite in many applications. The second aspect concerns the integration of complementary information. Most applications make use of VIR and SAR imagery. This trend continues to extend into hyperspectral images and other data, such as LiDAR. The latter provides topographical information and increases the discrimination of land cover types with similar spectral characteristics (Alonzo et al. 2014).

6.1 Urban Studies

According to a global study in 2014, 54% of the world's population live in urban areas, compared to 30% in 1950. The prognosis is that the percentage will increase to 66% in 2050 (UN 2014). There are already 34 mega-cities with more than 10 million inhabitants (Demographia 2015). The value for the total land mass covered by urban area lies between 0.5% and 4%. The literature does not provide an accurate measure, which urges the need for using remote sensing to accurately map urbanization and its processes (Angel et al. 2012). The aspect of sustainability in further development will therefore increasingly focus on cities in the future. The impact of rapid urbanization and industrialization can lead to environmental problems such as increased greenhouse gases emissions, groundwater depletion, surface temperature rise, and precipitation levels, just to name a few (Hu et al. 2015). The monitoring of the urbanization process plays a central role in urban pollution, energy consumption, and risk reduction in natural/anthropogenic hazards as well as climate change studies (Gamba 2014) and are summarized under "urban footprint" (Netzband and Jürgens 2010). Planning, maintenance, and resource management in urban environments require detailed and up-to-date information sources. Increased spatial resolution led to the possibility of mapping complex urban environments in great detail. Relevant urban topics include the study of heat islands, sprawl patterns, environmental quality, rainfall-runoff modeling, anthropogenic heat, and air pollution (Yan et al. 2015). Other research focuses on wide geographical area analysis, including global mapping of urban extents or alternatively on risk mapping to mitigate natural disasters (Salentinig and Gamba 2015). It is necessary to describe, monitor, and even simulate urbanization processes at different scales (Netzband and Jürgens 2010). We can distinguish two different types of urban studies:

1. Global urban extent mapping
2. Local planning, maintenance, and resource management

Depending on the focus, coarse spatial resolution satellite data with global coverage or very high-resolution data with enough spatial detail are used to reveal small differences. The complex nature of urban environments requires quite some complexity in processing remote sensing data. This is where RSIF contributes to obtain the best possible information from the imagery.

6.1.1 Global Urban Extent Mapping

With the background of the continuous increase in urbanization and population growth, it is important to be prepared and in the know about urban development. The main objective of global urban extent mapping is to increase awareness and help taking the right actions to ensure sustainability in the years to come (Angel et al. 2012). In urban mapping, each of the different sensors has advantages and disadvantages, which imposes the use of multi-sensor data and along with RSIF. Using SAR as all-weather satellite at coarse resolution, the total brightness relates to the density of built-up areas

(Salentinig and Gamba 2015). Therefore, multi-temporal images can lead to the analysis of urbanization processes at a global scale. For global mapping, the extracted information relates to a very coarse resolution with cell sizes of 500 m, like, for example, MODIS-based mapping (Schneider et al. 2009). Interesting combinations for global urban extent classification are MODIS with Defense Meteorological Satellite Program (DMSP) nighttime lights data set and gridded population data. Whenever ancillary data are integrated, higher levels of data fusion are applied to extract the most intelligent information in the process (Schneider et al. 2003).

6.1.2 Local Urban Mapping

Major information contributions originate from various high-resolution satellites, often complemented by multiresolution SAR data (Salentinig and Gamba 2015) and high-density LiDAR (Man et al. 2015; Zhou and Qiu 2015). The latter is gaining relevance with the use of not only the height information but also the LiDAR's intensity data (Yan et al. 2015). LiDAR data provide highly accurate structural data, which can support class separation using varying heights of buildings and roads (Jensen and Im 2007).

Urban landscape comprises a complex combination of both built-up and natural objects (Yan et al. 2015), challenging automated processing techniques. Urban areas cover a mixture of multiple spatial scales, different materials and objects, and two- and three-dimensional features (Salentinig and Gamba 2015). The derivation of a spectral signature for urban areas is therefore impossible even though hyperspectral remote sensing has improved the situation due to many narrow continuous spectral bands. In case of SAR, the constraints are the geometric distortions of elevated objects, such as houses, the "double bounce" or also called "corner reflector" effect (Figure 6.2), and the nonuniqueness of backscatter behavior between urban area and bare soil (Ferro et al. 2009). The urban complexity cannot be met by a single sensor (Ji and Jensen 1999). Consequently, RSIF is an important contribution to this type of applications since it provides a combination of complementary observations that facilitates object discrimination. Owing to the availability of very high-resolution sensors, submeter spatial resolution spaceborne sensor-based urban studies gained tremendous importance. In most of the studies, where fused images were used for urban land cover classification, the addition of texture features increased the accuracy of the results (Zhang 2001; Shaban and Dikshit 2002).

A practical example of an RSIF on urban land cover classification is a study on Ulaanbaatar in Mongolia. The team studied four algorithms, that is, WT, BT, Ehlers fusion, and PCA, whereby the latter was implemented as PCA on the multi-sensor VIR/SAR list of bands (layer stack). The study revealed that BT produces the best spatial separation of urban features. The resulting fused images are shown in Figure 6.3. But the best data combination for classification appeared to be the original VIR (QuickBird) and SAR (TerraSAR-X)

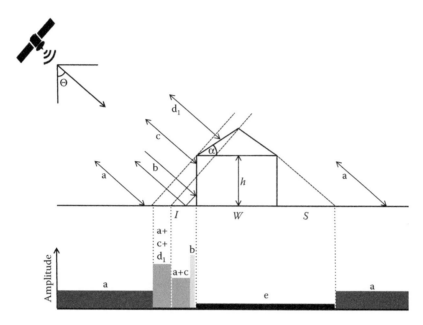

FIGURE 6.2
SAR double bounce or corner reflector effect from buildings. In the ground range image area, "a" is the intensity received of the scattering from the ground, "b" the double bounce, "c" the wall, "d" the roof, and "e" the shadow area. In the real world, *l* shows the area of layover, *w* the building width, and *s* the shadow area. d_1 indicates the backscatter from the roof facing the SAR sensor. A double-bounce effect can also occur with other elevated structures (e.g., trees). (Adapted from Ferro, A. et al. 2009. An advanced technique for building detection in VHR SAR images. Paper read at *Image and Signal Processing for Remote Sensing XV*, 31 August 2009, Berlin, Germany.)

bands along with texture derivatives of the data inserted into a Bayesian classification algorithm leading to 90.96% overall classification accuracy. The flowchart is available in Figure 6.4 (Amarsaikhan et al. 2010).

In a later study, using knowledge-based classification, the comparison of BT-, GS-, PCA-, and IHS-fused QuickBird imagery led to the best classification results for the GS fusion (Amarsaikhan et al. 2013, 2014).

Another group also compared pansharpening approaches, for example, PCA, subtraction, multiplication, BT, DWT, HPF, modified IHS, Ehlers, and HCS, to extract urban features. Their quality assessment resulted in identifying PCA and BT well suited for building, tree, and road identification. The statistical analysis revealed that HCS, subtraction, and HPF perform best (Dahiya et al. 2013). Ghanbari and Sahebi (2014) studied IHS, BT, synthetic variable ratio (SVR), and an improved IHS algorithm to fuse images of Shiraz city in Iran. The fused images are then classified to derive individual land cover classes. They identified their improved IHS algorithm considering local context as most suitable. Using PCA on optical (SPOT-5) and multi-polarized radar (Radarsat-2) data, Werner et al. (2014) classified urban

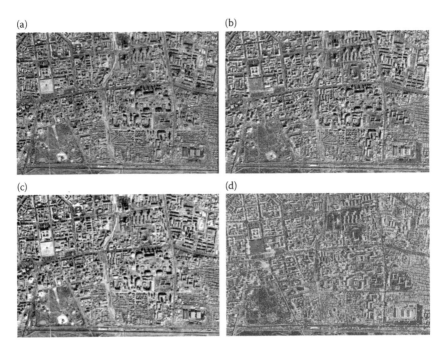

FIGURE 6.3
Comparison of fused images: (a) wavelet-based fusion; (b) Brovey-transformed image; (c) Ehlers fusion; and (d) PCA (red = PC_1, green = PC_2, blue = PC_3). (Adapted from Amarsaikhan, D. et al. 2010. *International Journal of Image and Data Fusion*, 1 (1): 83–97.)

land cover changes on Vancouver, Canada. PCA-fused data were compared to simply layer-stacking the multi-sensor bands. PCA (73.3%) performed worse than just using the stack of bands (89.1%) in the maximum likelihood classification.

6.1.3 Outlook

A new trend is the use of multi-angular images of high-resolution VIR sensors, such as WorldView-2. The multi-angle capability reduces revisit time and allows high off-nadir image acquisition. In the classification of urban features, there is a significant improvement in accuracy when applied to a multi-angle WorldView-2 data set (Longbotham et al. 2012). Classes, such as skyscrapers, bridges, high-volume highways, and parked cars are distinguishable. With the necessity of proper object discrimination and mapping, another trend leads from fusion at image level to feature level to take advantage of object-based analysis (Gamba 2014). Problems such as between-class spectral confusion, within-class spectral variation, shadowing, and relief displacement require higher-level fusion and the integration of other than image data (Berger et al. 2013a,b; Dahiya et al. 2013; Man et al. 2015; Yan et al. 2015).

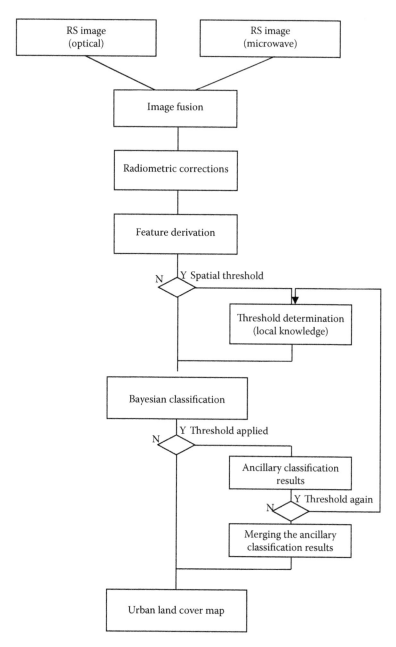

FIGURE 6.4
Flowchart of RSIF case study in Ulaanbaatar using Bayesian classification. (Adapted from Amarsaikhan, D. et al. 2010. *International Journal of Image and Data Fusion* 1 (1): 83–97.)

6.2 Agriculture

With the steady increase in population and the reduction in cultivatable land, food security becomes a central issue in agriculture. For the past five years, the contribution of agriculture to the gross domestic product (GDP) accounts for about 5%–10% in most regions of the world, while it is a major factor (up to about 60%) in Central Africa according to World Bank data (WorldBank 2015). There are only three options to increase land-based agricultural production: (1) expansion of farmland at the cost of other ecosystems, such as forests, (2) intensification of existing farmland, or (3) recultivation of already abandoned farmland (Stefanski et al. 2014). Remote sensing and RSIF are valuable tools in farmland mapping and crop monitoring. With the increase in spatial and spectral resolution remote sensing contributions to precision agriculture, a concept using observations of inter- and intra-field variability from different sources to feed a decision support system (DSS) are feasible. Mulla (2013) defines precision agriculture as a "… better management of farm inputs … by doing the right management practice at the right place and the right time." Precision agriculture is one of the strategies to respond to the threatened global food security (Gevaert et al. 2015). It leads to improved crop production and environmental conditions. Depending on the specific application within precision agriculture, spatial resolution requirements vary between 0.5 and 10 m (Mulla 2013), which can be delivered with today's satellite sensors in orbit. The spatial resolution and repeat cycle of sensor observations determine how much detail soil and plant characteristics can be mapped. Spectral, spatial, and temporal resolutions have increased during the past decade and therefore increased applications in agriculture and precision farming in particular. Parameters of interest include weed densities, crop height, leaf reflectance, nutrient, biomass, moisture status, soil properties and conditions, and other characteristics that are needed for fertilizer and pest management (Mulla 2013).

6.2.1 Mapping Growth Cycles, Plant Health, and Cropland Extent

An important RSIF application in agriculture is the mapping of crops for resource management in terms of crop types, vegetation health, and changes in cropland extent (Johnson et al. 2014; Stefanski et al. 2014). In VIR remote sensing images, agricultural land shows typical temporal patterns due to the phenology of planted crops and human activity (Stefanski et al. 2014). In order to derive useful parameters, it is necessary to acquire the data at certain points in time, which introduces the problem of cloud cover that is limiting image acquisition. Multi-temporal and multi-sensor image fusion including SAR and hyperspectral data is therefore a useful tool to overcome these limitations. RSIF comes in when spectral indices are created based on

TABLE 6.1

Selected Spectral Vegetation Indices Used in Agriculture

Index	Equation	Usefulness
NG	$G/(NIR + R + G)$	Carotenoids, anthocyanins, xanthophylls
NR	$R/(NIR + R + G)$	Chlorophyll
DVI	$NIR - R$	Soil reflectance
GDVI	$NIR - G$	Chlorophyll, N status
NDVI	$(NIR - R)/(NIR + R)$	Vegetation cover
GNDVI	$(NIR - G)/(NIR + G)$	Chlorophyll and photosynthesis, N status

Source: Adapted from Mulla, D. J. 2013. *Biosystems Engineering* 114 (4): 358–371.
Note: A = adapted, D = difference, G = green, N = normalized, NIR = near infrared, R = red, RVI = ratio vegetation index, VI = vegetation index.

ratios of reflectance values, which are very popular in agricultural monitoring (see Table 6.1).

Based on the ratios, leaf area index (LAI), biomass, chlorophyll, or nitrogen content can be assessed (Mulla 2013). Another successful VIR/SAR exploitation uses object-based classification and trajectory analysis to derive changes in cropland extent, mapping cropland from grassland at different points in time (Stefanski et al. 2014). With RSIF methods for pansharpening, such as IHS, BT, and AWT along with hybrid combinations, that is, IHS-BT, BT-AWT, and IHS-AWT, sugarcane was mapped in Nepal. The team applied pansharpened Landsat using band stacking and band averaging in the classification process. SVM appears to be the best classification algorithm in this context, identifying IHS-BT providing the highest classification accuracy for hybrid fusion, AWT delivering high accuracy among the individual methods. As layer stack, however, the three fused images IHS, BT, and AWT performed even better (Johnson et al. 2014).

Climate change, environmental changes, and anthropogenic activities have a strong influence on crop growth and harvest. Using remote sensing, it is possible to understand the processes, improve yield, and prevent shortage in crop supply. Crop monitoring requires information at field scales, sometimes even at higher resolutions to map subtle changes in the plants to discover possible anomalies due to fertilization, water supply, or disease. Applications use crop phenology metrics to optimize crop planting and harvesting dates as well as crop yield estimations and reporting. Another important parameter is the evapotranspiration (ET), which models the exchange of water vapor between Earth's surface and the atmosphere. Water plays an important role in agriculture. ET helps to optimize irrigation and crop health and yield. STARFM, which was originally developed to fuse surface reflectance, works on ET (Cammalleri et al. 2014). A single sensor alone cannot satisfy the ET requirements in terms of spatiotemporal resolution. Fusion of different data sources is necessary. MODIS TIR provides the moisture status, while Landsat provides the spatial detail. STARFM works particularly well if significant precipitation occurs between the Landsat image acquisitions (Gao et al. 2015).

In this context, the spectral–temporal response surface (STRS) should be mentioned as a method to exploit multi-temporal multi-sensor data for agricultural crops. All pixels are characterized by their multi-temporal intensity and waveband. The STRS is the interpolation on these data points (Viera et al. 2002) and results in a 4D (latitude, longitude, wavelength, date) data set (Gevaert et al. 2015). A further development is the spectro-temporal reflectance surface (STRS*), based on adjusted reflectance for sun and viewing angle (Villa et al. 2013). The actual fusion takes place in the downscaling step based on spatial unmixing (Zhukov et al. 1999; Amoros-Lopez et al. 2011). Strictly speaking, this step combines pixel and feature fusion levels because it uses classification of the HSR image prior to the fusion process. The fused pixels are a linear combination of the estimated LSR value weighted by the corresponding LSR class membership (Villa et al. 2013).

Making use of the state-of-the-art technology and algorithms, a research group successfully studied STRSs on a potato field in the Netherlands. With UAV hyperspectral data acquired along with Formosat-2 and *in situ* measurements based on a multispectral radiometer within the growing season, they were able to derive very accurate information on structure and biochemical parameters, such as chlorophyll. With an advancement of the method to produce the STRS, they introduced an *a priori* covariance between spectral bands of similar signatures for an improved Bayesian inference interpolation (Gevaert et al. 2015). This approach overcomes the limitation of the original STRS production, which does not consider physical characteristics of the reflectance spectra and ignores observation uncertainties. Landsat and MODIS data are suitable for winter wheat development mapping using STDFA for LAI extraction at high spatial and temporal resolutions (Wu et al. 2015).

6.2.2 Yield Prediction

Yield prediction has become relevant to investigate improvements in food production. Up-to-date information is important to food security and policy, price development, and foreign trade. Monitoring using remote sensing is indispensable due to the large coverage needed and the timeliness needed (Amorós-López et al. 2013). Another factor is the possibility to access crop and field information without destroying individual plants remotely. Remote sensing delivers possibilities to forecast yield and map protein content in cereal crops. The predication models are based on vegetation indices, such as NDVI, soil adjusted vegetation index (SAVI), difference vegetation index (DVI), and others, producing LAI, biomass, and nitrogen status estimates. Along with soil moisture changes, yield is projected (Doraiswamy et al. 2004; Wang et al. 2014). With increasing resolution, better crop estimates can be provided. Vegetation indices are used as indirect measures for crop yield (Yang et al. 2009). The combination of multiple sensors allows the identification of the best periods for grain yield estimations (Busetto et al. 2008; Wang et al. 2014). For yield estimation, the STARFM algorithm was converted into a spatial and temporal adaptive vegetation index fusion model (STAVFM). It combines NDVIs from low/high spatial and high/low temporal resolution images into a high spatial and temporal data set (Meng et al. 2013).

Fusing SPOT-5 and HJ-1A/B data over wheat fields in Henan and Jiangsu Province in China, researchers pinpointed initial gain filling and anthesis stages as best periods for estimating grain yield and protein content. They identified that the accumulated indices provided higher prediction accuracy (Wang et al. 2014).

6.3 Land Use/Land Cover Mapping

Knowledge about LULC at different scales is important to understand the dynamics of Earth processes and their influence on our environment and climate change. The distribution of land cover has a significant influence on Earth's radiation balance. Land cover is defined as compositions and characteristics of land elements on Earth's surface (Cihlar 2000). Anthropogenic land use and land cover changes have a strong impact on environmental change at local to global scales. Ecosystem health, water quality, and sustainable land management are significantly dependent on these impacts. Therefore, up-to-date global land cover information is important with special classes dedicated to urban, wetland, and forested areas (Friedl et al. 2010).

Depending on the scale and effort, land cover mapping relates to local, regional, national, or global studies. Multi-temporal remote sensing provides the means of the necessary coverage and repetition for monitoring LULC changes. These studies often cover several decades of development. Optical as well as radar data serve the purpose of identifying the developments. At global levels, it started in the 1980s with a very coarse spatial resolution (8 km) based on National Oceanic and Atmospheric Administration (NOAA) advanced very high-resolution radiometer (AVHRR) images. Later, the resolution increased to 1 km in the 1990s. New sensors purposely designed for global land cover mapping emerged (Landsat 7, SPOT-4 VEGETATION, MODIS, MERIS, and Global Imager—GLI) (Cihlar 2000).

With improved spatial resolution and regional adaptions of classification methods, LULC ambiguities at global level start to be resolved (Komp 2015). Regional factors have to be considered because land cover classes naturally differ from region to region as do seasonal windows. However, other problems are introduced, such as height differences, shadows and variance in illumination. High-resolution images have less homogeneous areas by nature. The combination of satellite resources is a crucial element to overcome limitations and fill gaps.

A pioneer in image mapping using RSIF was Chavez (1986). He fused digitized aerial photography and Landsat TM with selective PCA and different RGB displays for visual interpretation. In general, classification techniques are applied to derive land cover patterns based on spectral signatures and/or backscatter behavior and coherence on multi-sensor layer stacks. In the meantime, advanced methods have been developed to automate the process using, for example, automated processes (Bruzzone and Prieto 2000; Celik and Kai-Kuang 2011) and OBIA (de Almeida Furtado et al. 2015). LULC is very relevant to change detection leading to important input in environmental and socioeconomic studies. Fusing VIR and SAR images can improve classification results for large areas (Balik Sanli et al. 2009). VIR/SAR image fusion for mapping and map updating is particularly important in tropical areas where frequent cloud cover can cause problems in optical data acquisition. A very successful approach is the masking of clouds prior to image fusion (Pohl 1996). The best fusion technique for multi-sensor image maps using ERS-1, JERS-1, and SPOT XS above highly elevated terrain appeared to be multiplication (Pohl and Genderen 1999). However, the experiment was carried out long before more sophisticated and adaptive methods were developed.

Published studies show that different study sites (different vegetation structure, hydrological patterns, infrastructure, and urban makeups) require different approaches. In any case, the combination of optical and radar remote sensing data is recommended (Zeng et al. 2010). River floodplains benefit from multi-sensor classification based on simple layer stacks (de Almeida Furtado et al. 2015). The classification of grass-dominated wetlands seems to deliver higher-quality results when fused images are used (Li and Chen 2005; Castañeda and Ducrot 2009). The availability of multiple SAR data polarizations provides increased possibilities. An RSIF study for land cover classification used dual-polarized Radarsat and Japanese phased array L-band synthetic aperture radar (PALSAR) in combination with Landsat TM. The researchers identified DWT as the algorithm with the best overall results based on maximum likelihood (ML) classification (Lu et al. 2011). Other tested approaches included HPF, PCA, and normalized multiplication. An area in the Antarctic covered by snow, ice, rocks, lakes, and permafrost served a land cover classification study involving pansharpening prior to the mapping process. The team tested six pansharpening algorithms, that is, PCA, BT, GS, Ehlers fusion, and WT-PCA and created three SIRs, in which the pansharpened VIR images were processed (Jawak and Luis 2013). Using empirical thresholds, the final land cover layers were determined. Two separate quality assessments revealed that in terms of pansharpening and classification accuracy, GS and the hybrid approach of WT-PCA performed best to produce maps of snow/ice, water, and landmass of Antarctica. In a protected area mapping project in Uganda, the researchers fused TerraSAR-X and Landsat ETM+. They compared HPF, PCA, and hybrid WT-PCA fusion prior to classification. The hybrid fused data led to the best results (Otukei et al. 2015).

LULC mapping has already touched on some of the applications that will be discussed on in more detail in the coming sections. The next subchapter deals with change detection, again covering a large number of different applications due to the nature of the topic. The various applications will be further elaborated in the following sections.

6.4 Change Detection

The first definition of change detection in the context of remote sensing is given by Singh (1989) as "... the process of identifying differences in the state of an object or phenomenon by observing it at different times."In an update on the topic, Lu et al. (2014a) describe that "... change detection is a comprehensive procedure that requires careful consideration of many factors such as the nature of change detection problems, image preprocessing, selection of suitable variables and algorithms." That is why the matter is rather complex to be tackled in general. It is self-understood that the choice of appropriate dates (temporal resolution) for image acquisition is essential. The observation dates need to match the temporal interval of the change to be observed (Coppin et al. 2004). Another relevant aspect in this context is the ability to quantify the changes using data acquisition at different dates. Remote sensing-based change detection offers two primary categories: (1) bi-temporal (two dates, type, and extent) and (2) trajectory-based (three or more dates, multi-temporal, trends) change detection (McRoberts 2014).

The change detection procedure is divided into six phases (Lu et al. 2014a):

1. Identification of the change detection problem
2. Selection of appropriate remote sensing data
3. Image preprocessing, for example, RSIF
4. Choice of suitable variables
5. Application of suitable change detection (CD) algorithms, including RSIF
6. Quality assessment

Image and data fusion is recognized as a tool to integrate different remote sensing and other information to improve CD achievements, whereby pixel-based methods are common (Lu et al. 2014a). As depicted in Figure 6.5, RSIF is possible in step 3 to provide the best achievable preprocessed data to highlight and in step 4 to extract the changes, that is, the actual CD processing takes place. Similar to RSIF, change detection processing based on multi-source data requires accurate geometric rectification, conversion of data

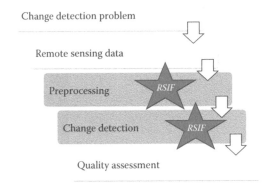

FIGURE 6.5
Position of RSIF in the LULC change detection process.

formats, and resampling. Therefore, the implementation of RSIF is straightforward and common in CD.

Change detection is in particular relevant to LULC observations with associated focus applications, such as deforestation and forest degradation, environmental pollution, hazard and disaster monitoring, urban change as well as crop stress, and health management. The literature distinguishes three different sets of categories (A, B, and C) of change detection. A1: change measurement and A2: classification followed by postclassification comparison (Malila 1980) or B1: comparison of independent classifications as opposed to B2: classification of a multi-temporal data set (Singh 1989). The actual CD technique for B1 is the so-called "post-classification comparison" or "delta classification," where two or more acquisitions are classified separately. B2 comprises the so-called "composite analysis," which can also be found as "spectral/temporal change classification, multi-date clustering or spectral change pattern analysis." This method carries the advantage of using the most original data, but it is rather complex (Coppin et al. 2004). Alternatively, we can distinguish C1: supervised and C2: unsupervised approaches. Supervised processing classifies the images based on ground truth or other addition information for the training of the classifier. Unsupervised techniques analyze the digital data without additional information (Bruzzone and Prieto 2000).

The differences to discover obviously depend on the specified application and thus influence image, band, and/or index selection. The success of change detection efforts mainly depend on the accuracy of image registration and calibration, ground truth availability, the knowledge about and complexity of the study area, change detection method, classification, and, of course, analyst skills, time, and cost restrictions (Lu et al. 2004b). For certain types of applications, the use of RSIF is described in the assigned sections of this chapter. Other applications and more general information on RSIF in CD are explained in this section.

6.4.1 Fusion Techniques in Change Detection

Univariate image differencing and ratioing, apparently the most popular CD methods, belong to the arithmetic group of RSIF techniques apart from multi-temporal color composites. They are relevant for the determination of change/nonchange questions. In addition, ratioing is essential for the production of indices and in geology. Its use is straightforward and enhances spectral reflectance differences. These enable mineral, soil, and vegetation discrimination (Langford 2015). PCA is popular in change detection as well (Deng et al. 2008). It requires a thorough analysis of the eigenvalues and a visual inspection of the data to deliver proper results (Coppin et al. 2004). Its advantage is that it can accommodate more than two images at once. In the context of CD, PCA is referred to as bi-temporal linear data transformation. Similar techniques are Tasseled Cap and GS with the advantage of reducing redundancy and enhancing differences between bands (Lu et al. 2004b). Another technique of CD is image regression, which statistically relates the two multi-temporal images through a stepwise regression. A hybrid approach was developed by Celik and Kai-Kuang (2011). It works on single-sensor multi-temporal images using RSIF, that is, image subtraction followed by an UDWT. The resulting fused data are then segmented to produce the change map. The advantage of this approach is the high reliability and the fact that it is an unsupervised approach. Another very interesting hybrid algorithm, developed by Khandelwal et al. (2013), uses two different CD methods, namely, differencing and CVA. The resulting images are fed into a DWT, resulting in a high-quality absolute difference (AD) image. Then, a neural network (NN) produces the final binary change map, containing the information of "changed" or "unchanged" areas. In comparison with PCA and NSCT CD, the hybrid technique resulted in the best percentage of correct classification (PCC) of 99.86%.

In multi-temporal CD, it is important to consider the impact of the presence of clouds, smoke, fire scars, seasonal flooding, and intensive agriculture in the interpretation of the fused images. These occurrences severely influence the results of CD analysis (Langford 2015). It is helpful to consider date, season, and characteristics of the objects investigated while selecting the dates and data.

6.4.2 Examples of Change Detection

Water quality monitoring is an essential application of CD. A recent study investigated the use of three sensors, that is, Landsat TM, ETM+, and MODIS, to observe the lake Albufera de Valencia in Spain. They applied STARFM to two pairs of input images even though the algorithm can also work on a single multi-sensor image pair (Doña et al. 2015). In coastal zone studies (Trabzon, Turkey) (Gungor and Akar 2010) and urban monitoring (Trento, Italy) (Bovolo et al. 2010), researchers successfully used WT for image fusion prior to CD processing. The Italian research team identified AWLP as the better performer in comparison to GIHS and GS in CD using CVA. However, in the end, the MMSE fusion appeared to have the best outcome in CD. The study revealed that CD results are scale dependent. They used a CD-specific evaluation method to rate five different fusion algorithms. Assessing the pure pansharpening performance with ERGAS, Q4, and SAM measures, MMSE and CBD fusion delivered the best results even though CBD performed worse for the CD capability analysis. This example shows the relevance of an appropriate quality assessment approach. The selection of an appropriate RSIF technique is a trade-off between the improvement of CD results and the introduction of artifacts from the fusion process. CBD can introduce artifacts locally. GS and GIHS over-enhance vegetated areas and therefore reduce their accuracy in CD (Bovolo et al. 2010).

An application in forest monitoring provides insight into the practical workflow in RSIF for CD. A research team studied Landsat-5 TM and SPOT-5 high-resolution geometric (HRG) image fusion to identify forest degradation and deforestation activities in the Brazilian Amazon (Lu et al. 2008). They transformed the registered and resampled TM image into principal components, normalized the HRG PAN to the first principal component (PC_1). Then, the result is used to replace PC_1 and a reverse PCA applied. The actual CD analysis is performed on the difference image resulting from image subtraction (fused image minus TM image). Strictly speaking, they followed a hybrid fusion approach using PCA and subtraction. The two-step analysis used the postclassification results (vegetation conversion information) and the hybrid fusion results (forest degradation and nonforest vegetation loss or gain) to provide a more sophisticated CD outcome.

6.5 Geology

Geology is a multidisciplinary application that studies composition, structure, and history of Earth, including its processes (van der Meer et al. 2012). Geological maps display the spatial distribution of patterns of geological units, structures, and rock properties. Remote sensing is a main contributor of information to mapping and updating because it allows a broader perspective and reveals relationships among strata and structures (Bahiru and Woldai 2016). The role of remote sensing is to provide information on geological settings, rock types, and for mineral exploration. They map faults and fractures and support the recognition of hydrothermally altered rocks based on spectral signatures and texture (Sabins 1999). Geology was one of the first applications to use RSIF and integrate VIR/SAR data. Pioneers of RSIF in geology used SPOT, Landsat, and Seasat to explore geological structures (Chavez et al. 1991; Yésou et al. 1993). The richness of interpreting a combination of multispectral information from optical sensors with surface roughness and texture from radar imagery has been enhanced with the availability of spaceborne sensors. VIR/SAR image fusion provides comprehensive geological information, for example, for metallic deposits even if they are obscured by dense vegetation in the tropics (Pour and Hashim 2014). By means of integrating multi-temporal images from optical sensors, it is possible to derive high-resolution mineral information even though the individual image does not have enough spectral resolution (Langford 2015). In geomorphological applications, VIR/SAR fusion is a recognized tool, in particular to map subsurface structures. Component substitution techniques enable the mapping of sand-buried arid landscapes (Rahman et al. 2010). Standard processing in geological remote sensing includes contrast stretching, band ratios, PCA, decorrelation stretch, and filtering along with IHS transform and HPF. Even though single-sensor color composites are not image fusion per definition, they contribute much useful information in geological applications and are therefore mentioned here as well. RGB composites, often also mentioned as false color composites (FCCs) of Landsat for lithological unit mapping use combinations of bands 731, 732, or 473. The latter is especially useful in densely vegetated areas. The band combinations benefit from strong absorption features of carbonate- and hydroxyl-bearing minerals. They reveal the spectral slope between 0.4 and 0.7 microns leading to iron-oxide minerals (Schetselaar et al. 2008). The PC_1 serves the identification of relief features to outline lithological units. Similarly shaded DEM data, for example, from SRTM, can be displayed with the RGB composites to facilitate tracing boundaries between quaternary units. The processing including RSIF enhances tones, hues, image texture, fracture patterns, and lineaments. The enhanced images are interpreted by visual interpretation and classification (Mwaniki et al. 2015). Pansharpening helps geologists to obtain more detailed maps of lineaments because the detection is directly related to spatial resolution. Hyperspectral remote sensing plays a more and more important role in geological and mineral exploration, since it helps lithological discrimination of rocks. In this context, PCA is used to reduce excessive data dimensionality (Kavak 2005).

From research in geological applications, the remote sensing community developed specific ratios that displayed in RGB to serve interpretation. It started with the availability of Landsat MSS data in the beginning of the 1980s (Goetz and Rowan 1981), evolved with SPOT due to its higher spatial resolution and stereo-viewing capability and became fully established with Landsat TM. The aim is to effectively identify and discriminate various materials. They are called inter-band ratios and are based on Landsat bands. For Regolith mapping, the ratios of band5/band7, band4/band7, and band4/band2 are fed into the red, green, and blue channels, respectively. In publications, it is noted as 5/7:4/7:4/2. Others suggest 5/7:5/4:4/1 color composites for the interpretation of lithological components (Kavak 2005). In addition, geologists use a decorrelation stretch of bands 754 (RGB), a variation in PCA. For D-stretching, three channels of multispectral data are transformed on

to principal component axes, stretched to give the data a spherical distribution in feature space and then transformed back onto the original axes.
Geologists found that ratios are most effective for the identification of surficial concentrations of minerals, such as hematite, goethite, and kaolinite
(Gozzard 2004). The benefit of band ratios is the fact that they eliminate
variations due to topography, reflectance, and brightness. Ratios in combination with multiplication can serve rock discrimination because these ratios
enhance specific chemical and mineralogical components of rocks (Langford
2015). An overview on a ratio interpretation scheme is provided by van der
Meer et al. (2012). The launch of the Terra satellite in 1999, with its 14-band
advanced spaceborne thermal emission and reflection radiometer (ASTER)
sensor, greatly enhanced mineral mapping. The sensor covers the visible
near infrared (VNIR), shortwave IR (SWIR), and TIR spectrum at 15 m, 30 m,
and 90 m spatial resolution, respectively. The geological community has
derived mineral indices based on ASTER data (Cudahy and Hewson 2002).
Hyperspectral remote sensing introduced a new dimension to quantify rock
or soil chemistry as well as physical characteristics through the spectra. A
main driver for hyperspectral data in geology is the study of epithermal gold
systems and mine tailings (van der Meer et al. 2012). Hydrothermal alteration mapping is used to indicate ore potential.

A comparison of different RSIF techniques, namely, PCA, IHS, Brovey,
SVR, and WT to enhance structural features for geological interpretation was
conducted to be able to extract surface and subsurface elements in an area of
Sudan. The research team found that PCA and IHS are suitable to maintain
the spectral content while enhancing subsurface features. The other techniques were not able to provide similar results (Rahman et al. 2010).

In a case study to update geological maps and identify areas of gold mineralization in Uganda, Landsat TM, ASTER data, shuttle radar topographic mission (SRTM), and airborne magnetic and gamma-ray spectrometry data were
processed. The RSIF techniques applied are D-stretch, PCA, and IHS plus a
series of other image enhancement procedures, such as stretching and filtering. An arithmetic combination fuses the spectrometry data with Landsat
TM enhanced the appearance of geochemical content in the lithological units
by texture. For the enhancement of structural information, IHS, DEM data,
or ASTER can be valuable (Schetselaar et al. 2008). The hybrid approach of
using IHS-PCA on ASTER is common, too. IHS provides sharper and clearer
lithological drainages and structural features (Bahiru and Woldai 2016).
Owing to the fact that geologists have good experiences with selective PCA
to extract maximum variance and decorrelated data, it is rather logical to use
PCA/IHS together to fuse optical and SAR data. An example is the study
in the Brazilian Amazon to detect geological structures. This study also
considered the integration of SAR and gamma-ray data using IHS (Teruiya
et al. 2008). They also found that sun illumination and structure directions
are crucial in the interpretation process, which is why the team applied an
arithmetic sum of the airborne SAR and PC_1 of Landsat TM (excluding the

thermal band). This new, already fused image was then inserted into the IHS transform as new intensity, hue was represented by the gamma-ray data and the saturation was given a fixed value. Therefore, the final hybrid approach included all three fusion methods. It provided excellent interpretation capabilities for topographic, vegetation, and lithological information extraction. The integration of gamma-ray, DEM, and Landsat TM data is received as the most successful to enhance geological image interpretation (Schetselaar et al. 2008). IHS has also been used for Landsat ETM+ and ERS-2 SAR fusion to identify Bauxite-mineralized zones in India (Patteti et al. 2014), similar to HSV-fused Landsat ETM+ with Radarsat-1 for spectral properties and morphotectonic features (Abdelkareem and El-Baz 2014). The latter identified the fused data set as the superior compared to the individual image interpretation. IHS was also applied to SPOT-5 data for the delineation of information for gold mineralization in Saudi Arabia, whereby the panchromatic band served spatial resolution enhancement and structural mapping purposes (Harbi and Madani 2013).

Ratios and PCA are the most established and popular methods in geology. PCA has shown its potential in structural and lineament mapping using multi-sensor and multi-polarized SAR. ERS-2 SAR combined with IRS-1C multispectral data using PCA serves the generation of color composites that allow a better interpretation due to greater tonal, textural, and brightness variations (Pal et al. 2007). Selective PCA produces ferric iron (PC_2 of Landsat TM 1 and 3) and hydroxyl images (PC_2 of TM bands 5 and 7) (Chavez Jr and Kwarteng 1989). Paganelli et al. (2003) investigated geological structures in the western plains of Alberta, Canada using PCA on four Radarsat-1 scenes with ascending and descending orbits. The research team applied a feature-oriented PC selection (FPCS) (Crosta and Moore 1989) after geocoding and speckle filtering. FPCS, also known as Crosta method (Aydal et al. 2007; Langford 2015) identifies the most suitable PCs to map the structures or other features of interest. It is based on the analysis of the eigenvector matrix and provides information on which PC contains spectral information about specific minerals. After processing the images, the actual geological lineament interpretation is done visually. In the above-mentioned study, PCs 2 and 3 offered most useful information because of the topographic-enhancement inherent in these Radarsat-1 PCs, whereby PC_2 displayed the structural information while PC_3 supplied the drainage patters (Paganelli et al. 2003).

Often the data are processed in multiple paths in parallel to provide the largest variety of data for interpretation. An example is the exploitation of Landsat-7 bands with the exception of the thermal band for structural mapping in Kenya. The study applied independent component analysis (ICA), a data mining technique, IHS of selected bands (5, 7, and 3), and PCA. The results (IC1, PC_5, and S) are displayed in RGB as color composites for interpretation. In addition, FPCS (PC_2, PC_3, and PC_5) and band ratioing (3/2:5/1:7/3) provide another two color composites. The panchromatic band is used for image sharpening using Ehlers fusion and filtered for fine lineament detection (Mwaniki et al. 2015).

The examples show the importance of proper band selection and their intelligent combination. Some authors call it the combinational optimization problem (Li and Wang 2015). Many researchers rely on the statistical approach of using PCA. Others apply application logic in identifying the wavelengths that best represent the material they are looking for. Other solutions include the use of the OIF or all of the above.

6.6 Vegetation Monitoring

The monitoring of ecosystem and natural resource management requires knowledge of vegetation changes at different scales over time. Remote sensing is well established in this context as information source. Changes in phenology, climate, anthropogenic activity, and processes over a long period of time can only be monitored using multi-sensor approaches. Vegetation monitoring observes plant communities and vegetation types, including grassland, forests, woodlands, and other natural vegetation. Parameters include composition, structure, and condition of the vegetation as well as vegetation dynamics (Hill 2005).

Spatial and temporal changes in vegetated surfaces are studied as land surface phenology (LSP) using remote sensing (de Beurs and Henebry 2004). It is distinguished from vegetation phenology (VP), which represents the timing of plant processes, that is, growing stages, such as flowering, productivity, and leaf senescence (Kuria et al. 2014). In that respect, remote sensors deliver data characterized by either high spatial resolution with a long time interval or low spatial resolution in short periods of time (Bhandari et al. 2012; Gao et al. 2015). RSIF can bridge this gap by combining high spatial with high temporal resolution data to synthesize a high spatial resolution image at the time required. This is also valid for enhancing vegetated areas for visual interpretation. In the context of vegetation mapping, the goal is to obtain the best possible RGB representation showing at-surface reflectance of the vegetation in the image. IHS transform with a hue adjustment increases vegetation visualization potential. In a case study, the derived IKONOS intensity channel was modified with the NIR channel and processed with the panchromatic band to account for the PAN channel shift toward the NIR part of the spectrum. The hue adjustment enforced a more natural appearance of trees and vegetated land cover (Malpica 2007).

6.6.1 Spatial and Temporal Data Fusion Approach

This group of techniques, abbreviated as STDFA (Wu et al. 2015), comprises several algorithms. The selection of the remote sensing data is a trade-off between spatial and temporal resolutions. Here, RSIF can solve the problem. Fusion techniques combine high temporal and/or spectral resolution images and high spatial resolution images to produce high temporal and/or spectral resolution of high spatial resolution images. Especially, in the context of carbon budget calculation, soil monitoring, and climate change due to land use/land cover changes, vegetation mapping is crucial. Other applications include growing cycles and phenology changes. Multispectral, multi-temporal images along with microwave remote sensing data provide valuable input to the calculations.

Using joint classification (Kuria et al. 2014) and image fusion (Lu et al. 2014b) leads to higher accuracy than using the single data alone. The LAI is a key parameter in ecosystem processes and considered the most important biophysical parameter in the context of vegetation characterization (Gray and Song 2012). It is defined as 50% of the total green leaf area per unit ground area (Chen and Black 1992). Here, remote sensing and RSIF is the only choice to map LAI continuously over large areas. Traditional pansharpening techniques such as IHS, PCA, and WT have limitations with increasing spatial resolution ratio between the input images. Recently, other more suitable techniques were developed. As depicted in Figure 6.6, there are two methods to provide this information using remote sensing: (1) inversion of canopy reflectance models, which does not always lead to a unique result and (2) building an empirical relationship (e.g., regression analysis) between vegetation indices (e.g., NDVI) and ground-based LAI. Even though inversion models are attractive due to their physical meaning, the parameterization is complicated and does not always lead to a unique result (Gray and Song 2012). Strictly speaking, spectral vegetation indices (SIVs) form a type of RSIF because they use ratios of image bands. Wu et al. (2015) provide an overview

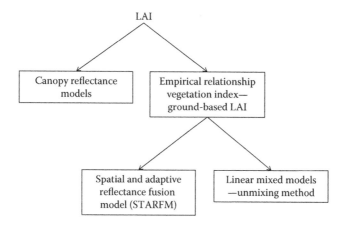

FIGURE 6.6
Methods to extract LAI from remote sensing data. (Adapted from Wu, M. Q. et al. 2015. *Computers and Electronics in Agriculture* 115:1–11.)

of eight vegetation indices used in vegetation mapping. The empirical relationship approach accommodates two categories of fusion algorithms: (a) the
STARFM and its enhanced version (ESTARFM) and (b) linear mixed models
(unmixing) (Wu et al. 2015). The advantage of using STARFM and its derivatives lies in the fact that it provides the vegetation indices frequently enough
to match the LAI measurement dates (Hankui et al. 2014).

Linear regression analysis in comparison to STDFAs led to the fact that
vegetation index fusion produces better results than reflectance fusion.
The usefulness of the EVI and NDVI to study photosynthetic activity and
morphology and therefore vegetation growth has been confirmed (Walker
et al. 2014; Wu et al. 2015). Testing the applicability of STARFM over dryland vegetation researchers found that the algorithm is sensitive to vegetation structure composition and reflectance characteristics of an individual
class. An example is the low correlation of the synthesized NIR band with a
reference Landsat-5 scene band 4 for all observed classes except grassland.
Furthermore, they found that it is not possible to compare results between
different ecosystem types (Walker et al. 2014). It is obvious that the higher the
spatial resolution of the data used in the reconstruction process, the better
the quality of the fused result and derived information (Hankui et al. 2014).

A long-term study using 38 Landsat TM and 115 MODIS images over a
period of 5 years proved that STARFM is useful to capture vegetation phenology in an area of *Eucalyptus* woodlands in Queensland, Australia. The
research team found a significant difference between the derived reflectance
of the original and synthesized images for different vegetation communities even though the pattern of means and standard deviations indicated the
good performance of STARFM. Rather high-resolution vegetation dynamics
mapping became possible. Applying ESTARFM and NDVI led to convincing results in semiarid areas even though the satellite images had a spatial
resolution ratio of 1:50. RapidEye and MODIS data provided accurate information in phases where vegetation growth was limited. The quality of the
results is very high and reliable since the calculated values were compared
to real high spatial resolution RapidEye images (Tewes et al. 2015). Others
found that introducing texture measures overcomes the limitations to produce SVIs with the empirical model. This is especially valid over forested
areas. They combined IKONOS (spatial information for texture) with Landsat
TM (spectral information for SVIs as predictors in the empirical model) and
MODIS (for the temporal information—phenology) using the parameters in
their empirical model (Gray and Song 2012).

6.6.2 Multimodal Classification

In the context of multi-sensor image classification, the improvement in classification accuracy obtained using RSIF is an accepted procedure. Unique
parameters of remote sensing images, for example, radiometric, spectral
(VIR), spatial and temporal resolutions, and polarization (SAR), deliver

relationships with certain types of vegetation and their conditions. In exploiting multispectral, radar, hyperspectral, multi-temporal, multi-angular images at different spatial resolutions, RSIF provides the necessary prerequisite. Apart from data selection to extract the desired vegetation parameters, the classification algorithm is of utmost importance. The classification relies on spectral signature, texture, and context information. RSIF is one tool to increase the quality of classification results by increasing the spatial resolution in multispectral imagery or introduce texture- and polarization-sensitive responses from SAR data. It is supposed to introduce the advantages of each single sensor in the fused product, reduce redundancy, and unmix mixed pixels (Hong et al. 2014). The combined use of VIR/SAR is valued in classification research. VIR and SAR images are combined to improve class discrimination, feature enhancement, and confusion reduction. Classification methods used to be overly applied on pixel basis. The trend leads to more advanced algorithms using OBIA or a combination of both. Very popular in vegetation mapping are

1. MLC because of its simplicity
2. SVM because of its good performance even if only very few training samples are available
3. CTA due to its robustness and capability to handle big and diverse data

The study of published research using RSIF for vegetation mapping reveals that the combination of VIR and SAR is essential and leads to higher classification accuracy. The variation in the different experiments lies in the form of combining or fusing the input data for classification. For the discrimination of grassland and alfalfa based on a MODIS/Radarsat-2 data set, a hybrid IHS–WT fusion prior to classification showed best results when compared to the single data sources. The case study resulted in 10%–20% better accuracy with fused images leading to better results than just layer stacking them (Hong et al. 2014). PCA, hybrid PCA-WT, HPF, and an arithmetic combination of VIR (Landsat) and SAR (ALOS PALSAR, Radarsat-2) data served vegetation classification in the Amazon in Brazil. The research team identified normalized SAR with TM data based on WT-fused images as best combination achieving an overall accuracy of 86.79% (Lu et al. 2014b). Another study investigated the benefit of RSIF for vegetation in wetlands in an area of the Amazon in Brazil. After a PCA on a Landsat TM scene, the first three principal components were fed into an IHS, where the intensity channel was replaced by a Radarsat-2 image after histogram match. In the OBIA segmentation following the fusion, the team compared the TM PCA classification with a pure SAR classification and a simple layer stack as input of the multi-sensor approach. They identified the layer stack of TM PCA and SAR as best input to the classification resulting in the least mistakes (Furtado et al. 2015). Wetlands represent complex land cover due to their heterogeneity of vegetation. They require advanced analysis methods. An example is the fusion of 3-band UAV VIR images with HH-polarized Spotlight TerraSAR-X images. After fusing the images for two dates using Ehlers fusion, which is very effective for VIR/SAR fusion, the research team applied spectral ratioing to remove illumination effects. Their classification approach followed the fuzzy MLC to allow the recognition of structures in their very high-resolution data set. The group achieved an 88.5% and 89.85% overall accuracy for two different seasons (Kuria et al. 2014).

6.7 Forest Monitoring

Forest plays an important role in regional and global carbon budget studies, climate change, and ecological modeling. Especially, owing to the extreme loss of forest cover for agricultural land, such as corn and soya in the United States and Europe and oil palm plantations in South East Asia, the focus has been driven to closely monitor deforestation. Recent studies revealed that in the period of 2000–2012, an area of about 2.3 million km^2 of forest was destroyed (Hansen et al. 2013). Large-scale mapping of forest, deforestation, and forest degradation studies heavily rely on remote sensing. RSIF has two functions: (1) provide information if certain data is not available and (2) increase the number of possible parameters to be derived from the data. The combination of active and passive sensor data lead to an increased reliability of the results and help the discrimination of different tree types (Sheldon et al. 2012). The use of radar remote sensing to complement VIR images in the tropics is indispensable. Pansharpening of single-platform sensors data helps to create high resolution and high accuracy of the classification results (Arenas-Castro et al. 2013). In the context of carbon stock estimation, tropical above-ground biomass receives increased attention. This is closely related to forest degradation and deforestation. VIR/SAR image fusion can contribute to the development of high-quality models if appropriate techniques are chosen (Basuki et al. 2013).

Stem volume estimations of forest stands in Sweden were investigated by Holmström and Fransson (2003), and they found that the results improved substantially when using optical and radar data in synergy. Goodenough et al. (2008) studied a very rich combination of different sensors in the field of forest analysis. The research benefited from the different and complementary characteristics of fully polarized ALOS PALSAR (forest classification), hyperspectral data (forest species and land cover), and LiDAR (vertical structure).

Multi-sensor data from optical and radar remote sensing satellites provide promising results in forest monitoring in the tropics. Demargne et al. (2001) use SPOT, ERS, and Radarsat SAR data to provide a forest inventory for Sarawak, Malaysia. They found SAR extremely useful in complementing missing information from the optical data. Dong et al. (2013) also implement information derived from VIR and SAR data for rubber plantation mapping. Their results can be transferred to oil palms. The authors are able to increase the classification accuracy to 96%. The use of multi-sensor VIR and SAR is very popular and successful for land cover mapping in the tropics. Erasmi and Twele (2009) found that multi-temporal SAR can be very helpful. However, the SAR data are not self-sustainable according to their opinion. The optimal approach uses the synergy of VIR and SAR data. For multimodal image processing, various analysis methods should be investigated among which joint classification of stacked multimodal images as well as fused remote sensing images are two options (Pohl 2014). The importance of multi-sensor data fusion is repeatedly mentioned. VIR and SAR are complementary data that provide a more complete perception of the objects observed. Of major interest still is the assessment of deforestation and oil palm plantation expansion using remote sensing. One large impact study was performed by SarVision in Sarawak covering the period from 2005 to 2010 using optical (Landsat) and radar (ALOS PALSAR) remote sensing (Wielaard 2011).

In a study of the Bogdkhan forest in central Mongolia, four image fusion techniques were explored, that is, modified IHS transformation, PCA, GS, and WT fusion (Amarsaikhan et al. 2015). Fused bands 2 to 5, and 7 of Landsat TM with C-band HH-polarized ENVISAT SAR imagery were used, both acquired within a few months of each other in 2010. After georeferencing of the optical and SAR images, using the topographic map at 1:50,000 and a forest taxonomy map, speckle suppression techniques were applied to help derive the SAR texture images. They found that the 3×3 Gamma MAP filter gave the best image in terms of delineation of different features as well as preserving the texture information. The result of the fusion exercise showed that the modified IHS transform and the wavelet-based fusion gave the best results in terms of the spatial and spectral separation of the different forest types. Figure 6.7 displays six fused images processed by the modified IHS, PCA, GS, CN, a WT-based method, and Ehlers fusion, respectively.

6.8 Natural Hazards and Disasters

The increased occurrence and dimension of hazards draws particular attention nowadays for various reasons. Population growth, climate change, and the ability to record, measure, and transmit information on hazards causes the importance of this application. Hazard is defined as a natural process or phenomenon that may cause a negative impact on society (UN-ISDR 2005). The group of natural hazards includes geological and meteorological phenomena such as volcanic eruption, earthquakes, coastal erosion, landslides, cyclonic storms (often referred to as typhoons or hurricanes), drought,

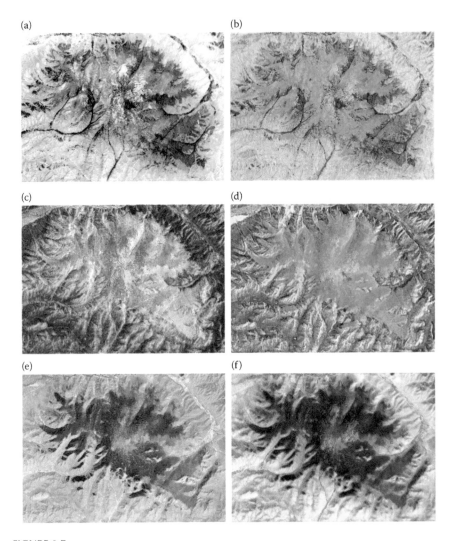

FIGURE 6.7
Fusion results of six different RSIF techniques for forest classification. (a) Modified IHS transformation; (b) PCA (red=PC1, green=PC3; blue=PC5); (c) Gram–Schmidt fusion; (d) color normalization spectral sharpening; (e) wavelet-based method; and (f) Ehlers fusion. (Adapted from Amarsaikhan, D. and N. Ganchuluun. 2015. *Image Fusion: Principles, Technology and Applications*, edited by C. T. Davis. New York, USA: Nova Science Publisher, pp. 83–121.)

diseases, and infestations as well as floods, storm surges, tsunamis, and wildfires (Alexander 1993; Tobin and Montz 1997; Smith and Petley 2009). They result from a mixture of climatic, geological, and hydrological factors. A hazard turns into a disaster once it causes great damage or loss of life. Remote sensing is an important contributor to a systematic framework for scientific knowledge of Earth and its processes, that is, for the development of a Digital

FIGURE 6.8

Five elements of disaster cycle. (Adapted from NASA [Green, D. 2015. Viewing the Earth's global environment from space: From scientific knowledge to social benefits. In *9th International Symposium on Digital Earth*, edited by Hugh Millward, Dirk Werle, and Danika van Proosdij. Halifax, Canada: IOP Conference Series: Earth and Environmental Science.])

Earth (Pohl and Genderen 2014). To protect life and predict potential natural hazards, information acquired using the available electromagnetic spectrum at high spectral and spatial resolution with a high repetition rate at global, regional, and local scale is of utmost relevance (Tralli et al. 2005). A good overview on the different spatial and temporal scales of selected natural hazards is provided by Gill and Malamud (2014).

In the context of managing disasters, there are five phases, that is, disaster, response, recovery, disaster mitigation, and pre-event preparedness, forming a cycle as depicted in Figure 6.8.

The monitoring of natural hazard or disasters benefits from high spatial and spectral resolution with a frequent revisit cycle (Green 2015). In the context of disaster prediction in the case of floods, typhoons, and volcanic eruptions, remote sensing also plays an important role. Therefore, RSIF to produce these integrated data sets at high spatial, spectral, and spatial resolution can play a key role. The information is needed for large areas in great detail (Nichol and Wong 2005). Based on the increased resolution by image fusion, spaceborne data can replace costly airborne surveys.

6.8.1 Floods

Flood monitoring using multi-sensor VIR/SAR images plays an important role. In general, there are two advantages to introduce SAR data in the fusion process with optical imagery. (1) SAR is sensitive to the dielectric constant, which is an indicator for the humidity of the soil. In addition, many SAR systems provide images in which water can be clearly distinguished from land.

FIGURE 6.9
VIR/SAR fusion for flood mapping; upper left: ERS-1 SAR postflood (inverted), middle: VIR/SAR fusion, lower right: Landsat TM bands 1 and 4 preflood.

(2) SAR data are available at any time of the day or year independent from cloud cover or daylight. This makes it a valuable data source in the context of regular temporal data acquisition necessary for monitoring purposes. For the representation of the preflood situation, the optical data provide a good basis. The VIR image represents the land use and the water bodies before flooding. Then, SAR data acquisition at the time of the flood can be used to identify flood extent and damage. An example is the fusion of Landsat TM and ERS-1 data over an area in the Netherlands where the rivers Rhine, Waal, and Maas severely flooded a central part of Holland. The flood impact and extent was mapped using TM bands 1 and 4 (preflood image) along with the SAR (flood image) in an RGB layer stack. In order to enhance the flooded areas, the SAR image was threshold stretched and inverted (Pohl et al. 1995). Permanent water bodies are shown in dark blue while flooded areas appear in light blue as depicted in Figure 6.9.

6.8.2 Volcano Eruptions

Remote sensing is a key contributor to operationally forecast, monitor, and manage volcanic hazards (Dean et al. 2002). Early detection of volcanic activity is crucial in the decision making and potential volcano hazard mitigation. For volcano monitoring, SAR interferometry (InSAR) has become a widely used technique, because of the many suitable characteristics of SAR satellites such as (i) day-and-night acquisition, (ii) all-weather capability, (iii) wide area coverage, and (iv) its ability to detect small displacements. However, during an emergency, when the volcano shows imminent signs of an eruption, it is difficult to use only one SAR satellite because of limited revisit times. TIR, VIR, and SAR data deliver information on ash clouds, gas emissions, and volcano activity. Temporal fusion methods such as multiple SAR satellite imagery can be used for monitoring purposes to detect changes. A good example of this is the study by Sango et al. (2015), where they fuse three SAR satellite data for dynamic monitoring. They used ALOS-2 (L-band), Radarsat-2 (C-band), and TerraSAR-X (X-band). Thus, such an approach of combining different SAR satellites for active monitoring provides more information than using only one type of SAR satellite. Another example of multi-temporal products from radar sensors to allow operational volcano monitoring and feed information into decision support systems followed by advanced interpretation techniques is given by Meyer et al. (2015). Automated products from ERS-2, Radarsat-1, and ALOS PALSAR complement AVHRR hazard information once every 5–10 days. These are (i) geocoded and radiometrically terrain-corrected amplitude images, (ii) amplitude change detection maps, (iii) differential InSAR image time series, and (iv) coherence information. Amplitude change detection supports the identification of an expected eruption, tracks the progress of an ongoing eruption, and maps lahar and lava flows as well as the ash fall. InSAR provides the necessary deformation information while coherence complements surface deformation signals prior to a volcanic eruption, similar to posteruption deposit assessment. ASTER multispectral images are utilized to complement the monitoring process (Meyer et al. 2015).

6.8.3 Landslides

Landslides are primarily caused by heavy rainfall, often resulting from typhoons and powerful rainstorms, and by other natural hazards, such as earthquakes. Landslides are relatively straightforward to detect on optical remote sensing imagery, in that they all normally have three common characteristics that are identifiable by visual interpretation. These are the point of origin defined by the rupture surface, a landslide trail along the slope, and a deposition fan where the greater part of the mass is deposited. A comparison of various image fusion techniques to compile a landslide inventory of an area in Brazil using Landsat and SPOT imagery identified IHS as the most suitable (Marcelino et al. 2009). The analysis considered five different fusion algorithms, namely, Brovey transformation, HSV, IHS, PCA, and wavelet transform. IHS transform, using the 10-meter SPOT PAN channel fused with channels 4, 5, and 3 from Landsat delivered the best result. Furthermore, it was shown that the capacity to identify landslides is directly related to their spectral response, their typical features (relation of length to width), the existing contrast with neighboring targets, and the geomorphological features of the area.

Landslide probability analysis is another important application of image and data fusion. It requires data acquisition over the same scene collected from different sources, for example, optical imagery, NDVI, and DTM data sets (Chang et al. 2014). The type of processing is another research focus in the context of landslide applications. Generalized Boolean functions aid the exploitation of multisource data fusion in landslide classification (Yang-Lang et al. 2007). Recently, various fusion algorithms were applied to classify and estimate the potential risks of landslides, and subsequently improve the ability to predict landslide risk trends, and thus prevent and minimize impacts (Chang et al. 2014). Multisource data sets are fed into the proposed band generation process (BGP) to generate and fuse additional bands for more distinct characteristic information, especially for landslides, originating from the multisource data sets. The BGP output is then fused through the Fisher criterion-based nearest feature space (FCNFS) for land cover classification, and the mapping results are visualized. In order for the FCNFS to be effective on multispectral images, they state that the number of bands must not be less than the number of materials to be classified. This ensures that there are adequate dimensions to accommodate the FCNFS resulting from the individual features. Such a degradation of classification performance may incur on multispectral images, as it is the case for the three-band SPOT images, where only a few feature bands are collected for classification. They therefore increased the three multispectral SPOT image bands by adding two NDVI channels plus a DTM channel. This larger number of bands maximizes the classification effectiveness of FCNFS as it also takes into account the features from the additional new channels, such as the slope information derived from the DTM.

One of the problems with detecting and mapping landslides from satellite imagery is the fact that many landslides are quite small in aerial extent, often less than 100 meters by 100 meters. Hence, fusing very high-resolution imagery with multispectral images enhances their detection capabilities (Santurri et al. 2010). Coupling both image fusion methods with change detection techniques is another approach that has proven useful to inventory landslides using satellite data (Nichol and Wong 2005).

6.9 Coastal Zone Monitoring

The monitoring of changes in coastal zones is essential to understand environmental impacts of human activities and natural processes (Chen 1998). High-quality geospatial coastal information is essential for coastal resource management, coastal environmental protection, and coastal development and planning. Acquiring such information has traditionally been carried out using aerial photographs and ground surveying. While these techniques have many advantages, such as flexible scheduling, easy-to-change configuration, and high-quality mapping results, they are expensive and need special logistics and processing procedures. Hence, with the advent of higher-resolution satellite data becoming available in visible, infrared, and microwave regions of the spectrum, much work has now been done on integrated coastal zone management using Earth observation data, especially using change detection and image fusion techniques. There have been many studies done in coastal zones using optical data as well as those using SAR data. Owing to its continuous availability, Landsat is a good system to perform multi-temporal monitoring. An example application is the mapping of coral reefs over a period from 1984 to 1997 using five Landsat TM scenes (Pitts et al. 2004). Another popular application is shoreline mapping (Chang et al. 1999) or coastal landscape change mapping (Drzyzga and Arbogast 2002). High-resolution IKONOS images served to monitor and map coastal landscape changes during spring and late summer seasons for three sites along Lake Michigan and one site along Lake Huron. A good example of an SAR coastal zone survey is that carried out by Tuell (1998), who used SAR for shoreline mapping and change detection in a remote area along the Alaska shoreline as part of the National Geodetic Survey (NGS) effort to evaluate the potential of several mapping technologies. Another active sensor used for coastal zone research is that of airborne LiDAR. LiDAR is of increasing importance for shoreline mapping and change detection (Cook 2003).

However, it was with the introduction of image fusion technologies that coastal zone studies received a major boost. The integration of active remote sensing technologies with other data sets made it possible to provide reliable and automatic solutions for coastal mapping and change detection. Some of the earliest examples of image and data fusion for coastal zone studies used LiDAR data and historical aerial photographs to study the Gulf of Mexico shoreline changes (Gibeaut et al. 2000), while others combined LiDAR data and IKONOS images for coastal mapping (Lee and Shan 2003). These results showed that merging these data sets greatly improved the quality of the classification process. It is also valuable to fuse hyperspectral AVIRIS imagery with LiDAR-based DEMs for detailed coastal mapping (Elaksher 2008). The results demonstrate that integrating laser and optical data can provide high-quality coastal information.

Coastal ecosystems are under threat due to climate change and anthropogenic activities. Their monitoring and management is important to preserve and restore the habitats for coral reefs, mangroves, and other structures, protecting the coast and influencing environmental health. These complex structures and processes need sophisticated data integration and hybrid fusion strategies (Henderson et al. 1998). Pansharpening techniques support the delineation of coastlines using the normalized difference water index (NDWI) (Maglione et al. 2015).

Tidal flats in many coastal zone regions of the world are a highly dynamic and largely natural ecosystem with high economic and ecological value but which are also at risk due to climate change, rising sea levels, algae blooms, invasive species, and marine pollution (Jung et al. 2015). There is a need for the detection of emerging changes and the potential loss of the natural or seminatural ecosystems accompanied by a decrease in water quality. Accessibility from both sea and land in such tidal flats is typically very poor, which makes the monitoring and mapping of tidal flat environments from *in situ* measurements very difficult. Hence, a multi-sensor image and data fusion approach is the optimum way to study such intertidal coastal zones.

Jung et al. (2015) used a combined multisensory analysis involving RapidEye and TerraSAR-X satellite data combined with ancillary vector data about the distribution of vegetation, shellfish beds, and sediments for the accuracy assessment. Although not strictly speaking an image fusion approach, they did multisensory data integration to show the value of such an approach. For example, the water coverage was separated from the tidal flats using NDWI. The shellfish beds were estimated with the textural features of the TerraSAR-X data and morphologic filters (MFs). Third, the classification of vegetation (saltmarsh, sea grass/algae) was based on the modified soil-adjusted vegetation index (MSAVI), object-based features, and exclusionary criteria. They then separated the remaining areas into different sediment types with an algorithm that uses a thresholding technique applied to radiometric values, the MSAVI, and a majority filter. Their results showed that they were able to identify the location and shape of salt marsh and shellfish beds using such multi-sensor remote sensing data. Their results also emphasized that a sediment-type classification could not be achieved with high accuracy using spectral properties alone due to their similarity, and hence using multi-sensors with different spectral, spatial characteristics was required.

An excellent report on the integration of multiple data sources to study coastal zones is that produced by Paine et al. (2013). In their detailed report, they integrated the use of the following data sources to determine wetlands distribution and coastal topography in an area of Alaska:

1. Color infrared aerial photography with 50 cm spatial resolution, acquired in 2012 at a scale of 1:10,000
2. SPOT RGB imager at 2.5 m spatial resolution acquired in 2009 and 2010
3. Landsat ETM+ imagery (multi-date)
4. The National Land Cover Data Base of the area, produced in 2001
5. The National Wetlands Inventory Maps, produced in the 1970s and 1980s from 1:24,000 aerial photography
6. Airborne LiDAR survey flights, flown in 2012

No single source was able to produce the required information; however, using these multiple data sources, a very detailed understanding was acquired of this sensitive coastal region.

6.10 Coal Fire Monitoring

Underground coal fires are not only a major environmental hazard, contributing greatly to CO_2 emissions worldwide, but they also cause a huge economic loss in lost coal resources. They are the reason for land degradation (Stracher et al. 2013; Song and Kuenzer 2014).

In addition to the loss of coal that is actually burnt, access to remaining reserves is often made difficult or impossible by the fires, and production suffers. Fires occur in most coal-producing countries, such as China, India, Australia, and South Africa (Gens 2014). Extraction of coal from the subsurface, without filling the empty spaces, has caused widespread land subsidence, and left the coal mine areas prone to subsidence, even after the mining activities have ceased. Such subsidence damages the infrastructure both above and near the subsided area, thus blocking the coal reserves. In addition to the environmental damage caused by the land subsidence, it has cost the lives of mine workers and people living above such areas.

Geo-environmental indicators of coal fires can be identified, mapped, and quantified by remote sensing techniques. Other coal fire parameters, including fire depth, can be indirectly inferred. In any case, it requires a multi-sensor approach (Prakash and Gens 2011). In China, for example, remote sensing satellite and airborne data studies have shown that coal fires occur in a belt of 5000 kilometers from north-west to north-east China, and 750 kilometers in north-south direction. What the many studies on coal fires in China and other countries have shown is that several different satellite and airborne sensors can partially detect such fires, depending on certain conditions, such as size of the fire, depth of the fire, terrain relief, time of image acquisition, prevailing weather conditions, and season. Hence, realizing that each sensor and its associated image-data type only provides part of the detection of the coal fires, remote sensing image fusion has become a valuable method to have a more robust and reliable result on the detection, location, size, and depth of the fires.

Typical data sources used are optical color and color infrared airborne data, TIR airborne data, plus satellite thermal and optical data, along with microwave satellite data. For example, optical images show the burnt rock above a coal fire quite distinctly, but does not indicate whether the coal fire underneath is still burning or it is a paleo fire that are known to have occurred since Pleistocene times. Similarly, TIR data can identify hot spots that could be caused by underground fires, but only under certain conditions. Hence, multi-sensor image fusion results in more reliable data on the coal fires than would be possible using the images separately.

One of the first studies on RSIF to tackle this problem was that reported by Zhang et al. (1998). In their work, they describe a remote sensing method based on multi-sensor data fusion, consisting of fusing a variety of satellite-based image types (optical, thermal and microwave), together with airborne data (optical and thermal infrared imagery), as well as ancillary data sources such as geological and topographic maps. They used pixel-based, feature-based, and decision-based fusion approaches to detect, measure, and monitor the underground coal fires in the far north-west of China's Junghar coal basin. By combining optical satellite imagery with thermal infrared night time images, plus SAR images to make interferograms for monitoring the land subsidence, the fused imagery gives valuable insights into the affected areas and processes taking place. Using this approach, the authors developed an overall, integrated multi-sensor data fusion system, as depicted in Figure 6.10. This three-level RSIF uses very cheap NOAA-AVHRR data, which provide daily global coverage, together with the VEGETATION data on board the SPOT satellite. In addition, other thermal satellite data were used. From this broad-scale data, more detailed information is provided from Landsat thermal images, both day time and night time data, and ASTER TIR images, to confirm the areas detected in Level 1. At the third level of the fusion process, detailed airborne optical and TIR imagery, together with ancillary data such as field measurements and land subsidence measurements from InSAR, are used to make an overall classification of the size, depth, and direction of burning of the coal fires.

The fusion approach was further developed and described by more research (Prakash et al. 2001). The team studied the Ruqigou coalfield in

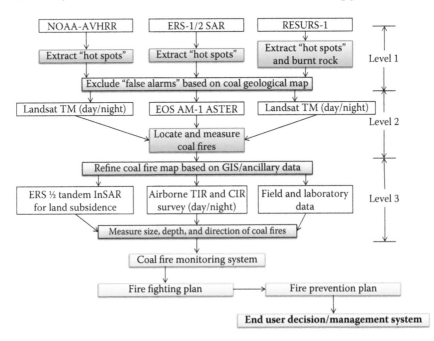

FIGURE 6.10
Three-level multi-sensor fusion system for underground coal fire monitoring. (Adapted from Zhang, X. M. et al. 1998. *Geologie en Mijnbouw* 77 (2):117–128.)

Ningxia Province, an area that produces high-quality anthracite coal for export. They used the following data sets in their comprehensive fusion work:

1. Landsat TM daytime image, for location of the mining areas, coal outcrops, and coal dumps
2. Landsat TM thermal infrared (TM Band 6) nighttime image, to detect the coal fires
3. ERS-1 SAR imagery to generate the interferometric products for locating the subsidence areas
4. A 1:5,000 large-scale topographic map with contour interval of 5 meters for the production of a digital terrain model (DTM) at 5-meter grid cells
5. Field measurements, including high-precision GPS measurements, thermal profiles along sections of interest, field photographs to serve as ground checks, and validation

Using these data sources, a decision-based fusion process was used, whereby all the individual image and data sources were all classified prior to the fusion process. This approach was adopted to obtain an overview of the problems in this important coal mining area, and to see the relation among mining areas, coal fires, and subsidence areas. Hence, the classified images generated from processing the optical, thermal, and microwave data were used as inputs. An FCC was applied to present these three different pieces of information in the three color channels. The coal fires, identified from the thermal data, were assigned to the red channel, the range change image showing the subsidence areas was displayed in green, and the coherence image from the interferometric SAR data was displayed in blue. They further enhanced their final product by merging it with a linearly stretched near infrared TM Band 4 in such a way that all the dark pixels on the FCC were replaced with this optical data. The result is shown in Figure 6.11.

Another interesting application of RSIF for coal fire detection was carried out by Kuenzer et al. (2007). They combined the multispectral surface-extraction algorithm for coal fire risk area delineation with the thermal anomaly extraction algorithm, to automatically detect coal fires from remote sensing data. Their results led to the creation of an operational detection and monitoring system for coal fires in large areas.

Shupeng and Genderen (2008) show some examples of how to fuse imagery with other data. Figure 6.12 shows a predrawn TIR night time image of a coal basin in China, fused with a DTM, created from color infrared stereo aerial photographs. The combination creates a 3D perspective view of the Kelazha anticline in Xingjiang Province of N.W. China. Such fused products assist in the study of underground fires, as they show the influence of topographic variables on the distribution of the fires. This enhances the location

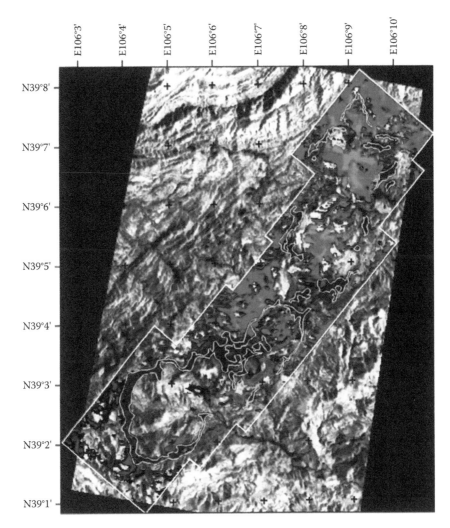

FIGURE 6.11
Final fusion product with information from optical, thermal, and microwave regions. (Adapted from Prakash, A. et al. 2001. *International Journal of Remote Sensing* 22 (6):921–932.)

and understanding of where the coal fires occur along the anticline. As depicted in Figure 6.12, a thermal infrared night time image of a coal basin in China is fused with a DTM. This enhances the location and understanding of where the coal fires occur along the anticline.

A recent study in 2015 reports on an application of RSIF for detecting surface coal fires in India (Roy et al. 2015). The team shows how problematic it can be to use two different data sets, that is, thermal and optical bands of the ASTER satellite and night time thermal images from Landsat TM. Factors

FIGURE 6.12
3D model of coal fires in the Kelazha anticline in Xinjiang, N.W. China, with a thermal infrared image draped over a DEM of the area. (Adapted from Shupeng, C. and J. L. van Genderen. 2008. *International Journal of Digital Earth* 1 (1):43–65.)

such as season, the chosen cut-off temperature, or day time or night time data all affected the identification ability of the fire zones. Seasonal effects were the main problem, as seasonal temperature variations have a pronounced effect on the temperature of background elements such as rocky outcrops.

What many studies on RSIF of underground coal fires have shown is that there are many benefits of a multi-sensor image and data fusion system for the detection, measurement, and monitoring of underground coal fires. They are summarized in Table 6.2.

6.11 Detection of Landmines and Minefields

Landmines are quite different for most other military weapons, as once they have been laid, they have the potential to maim or kill innocent civilians long after the conflict is over. Although the use of specific weapons, such as landmines, is restricted and regulated by international law, such as the Ottawa Convention, signed in 1997, many of the largest military powers such as the United States have still not ratified the treaty. There are still millions of landmines located in some 60 countries around the world. To detect and clear more than 600 different types of landmines is clearly a major problem. The difficulty of locating buried landmines is a problem that until recently defied an easy solution. Most of the landmines have been manually placed on or just below the surface, but nowadays rockets and aircraft may place them as well. These are even more difficult to find, as no precise map of the minefield

TABLE 6.2

Results and Benefits of Multi-Sensor Data Fusion for the Detection of Underground
Coal Fires

Results	General Advantages	Operational Benefits
Robust operational performance	Multiple sensors provide coverage even if a single sensor does not acquire data	• Allows continued operation • Increases possibility of detection
Enhanced spatial resolution	Pansharpening provides higher spatial resolution	• Aids interpretation and detection
Enhanced temporal coverage	Multiple platforms reduce revisit time and increase potential of image acquisition	• Day/night, all-weather capability • Increased possibility of detection
Enhanced spectral coverage	Complementarity of VIR/SAR/TIR increases detection capability; increased robustness	• Increases classification accuracy • Allows continuous operation
Increased confidence	Each sensor type/image helps to confirm detection; reduced number of false alarms	• Increased classification accuracy
Improved detection	Multi-sensor data fusion increases detection	• More reliable • Higher accuracy
Reduced ambiguity	Joint information from multiple sources reduce a number of alternative hypothesis whether a feature is a coal fire or not	• Faster decision making • Increased reliability of results

Source: Adapted from Zhang, X. M. et al. 1998. *Geologie en Mijnbouw* 77 (2):117–128.

is made. Minefields laid by military forces are usually laid in patterns, and their locations are mapped. However, over the years, since the mines were laid, the maps of the minefields have been lost or destroyed. Moreover, those mines laid by irregular military troops, such as freedom fighters during the wars of independence from colonial masters, or terrorist groups, are used as weapons of terror, aimed specifically at the civilian population, and hence no locational information is recorded. Such use of landmines typically aim to deny access to agricultural land, villages, water sources, along roads, footpaths, etc. in order to destabilize the economy, destroy food resources, and render the land useless until it is cleared.

   To locate landmines by traditional techniques such as handheld metal detectors, is slow, dangerous work, and requires much manpower. The integrated civilian approach to the landmine clearance problem is called mine action. It includes things such as mine awareness, risk-reduction education, minefield survey, mapping, marking, and clearance. Civilian demining can only start after the conflict has stopped, and aims to return previously mined land to its prewar condition. Hence, there is a growing recognition that remote sensing methods, both airborne and from satellite, especially when coupled with multi-sensor data fusion, can speed up the process of locating and clearing the minefields and individual landmines. Scheerer (1997) used state-of-the-art electro-optical, millimeter-wave, and radar sensors to detect landmines of different origin. An important distinction he makes is the difference between military demining and humanitarian demining. The military need a system that is fast, round-the-clock, and all-weather capability. With humanitarian demining, we have more time, but both must be reliable to avoid casualties.

During the 1980s and 1990s, the military began research on image and data fusion and applied remote sensing image fusion to the problem of detecting landmines and minefields, especially as no individual sensor is able to detect all types of mines in all types of conditions. Fusion algorithms and fusion architectures at sensor and information level have been studied (Breuers et al. 1999). The earliest work on remote sensing image fusion for civilian demining purposes was that carried out in the late 1990s by a consortium of European universities and companies. They carried out research and practical real-world RSIF projects in Europe, Mozambique, and Angola, using both airborne remote sensing platforms and satellite imagery (ITC 1999; Maathuis 2003). Playle (2006) fused bands of hyperspectral data in the United Kingdom, using both airborne and *in situ* sensor acquisition systems. They show that using very narrow spectral bands, there is more chance of detecting different types of surface-laid landmines. However, this method has only been shown to work in experimental studies. In real-world situation of minefields in developing countries with so many different types of landmines, so many different types of environmental and terrain conditions, it is not yet an operational technique.

A review on achievements in landmine detection by airborne and satellite platforms concludes that no single sensor can deliver the necessary information under all circumstances (Maathuis and Genderen 2004b). In a case study, RSIF of satellite images served the detection and mapping of minefields along the Mozambique–Zimbabwe border. The data set comprised aerial photographs from 1974 (during the early construction phase of this defensive minefield), KFA-1000 satellite image from 1979 (at approximately 5 meter spatial resolution taken during the conflict), 1:25,000 panchromatic stereo aerial photos from 1981 and 1989 (after the conflict), and a Landsat TM image from 1992 (Maathuis and Genderen 2004a). Fusing these temporal data sets shows the overall developments that have taken place along the border. Figure 6.13 displays the result. On the left is the old panchromatic aerial photograph, while the remaining part of the figure shows the fused result of Landsat TM image (bands 2, 3, and 4) with the KFA high-resolution image, using the IHS transform, where the intensity component was substituted by the high-resolution KFA image. The paper describes the benefits of using such a temporal remote sensing image fusion approach to delineate this minefield. These included the use of very high-resolution stereo aerial photographs to enhance the fine spatial details associated with

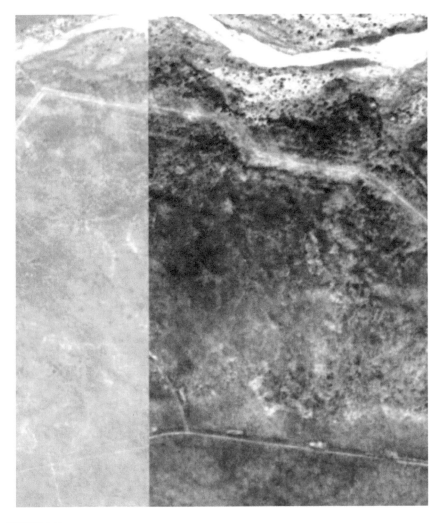

FIGURE 6.13
RSIF to monitor a border minefield in NE Zimbabwe; old aerial photography (left); and
IHS fused satellite images. (Adapted from Maathuis, B. H. P. and J. L. van Genderen. 2004a.
International Journal of Remote Sensing 25 (23):5195–5245.)

laying a minefield (roads, tracks, fences, etc.) plus change detection, to highlight the differences in vegetation between that existing before the laying of
the landmines, during, and after the completion of the minefield. By fusing
these temporal images, such fine changes in the land surface were greatly
enhanced and became much more visible than possible from any one single
image. This is one of the few published studies using pixel-based image
fusion to detect minefields.

In addition to the border example shown in the figure, other examples
utilized remote sensing images of minefields in the Netherlands, the Berlin

Wall, and the Iron Curtain constructed on the East Germany side of the former border West and East Germany. In image processing and interpretation, it is important to consider information on the many indicators or clues for detecting individual landmines and minefields (Maathuis 2008). The success of exploiting multi-sensor satellite images strongly depends on indirect indicators since individual mines cannot be identified for reasons of land cover and terrain conditions. With the indicators, however, it is possible to identify minefields (Maathuis 2003).

Another research group (Bloch et al. 2007) describes a multi-sensor data fusion approach for both airborne and spaceborne imagery to reduce the mine-suspect areas. For best results, they show that decision-based fusion, with the final decision being taken by the deminer, provided best results. Further studies revealed that multi-sensor data fusion potentials for both close-in detection and mine area reduction take advantage of all levels of fusion. For the remote sensing example, they used synthetic aperture radar and multispectral sensors. They showed how ancillary data are vital to the successful fusing of such images (Milisavljevic and Bloch 2008; Milisavljević and Bloch 2010). In this context, fusion is not an easy task, as the various remote sensors used may reside on different platforms (vehicle, UAV, helicopter, aircraft, satellite), leading to additional difficulties with image registration. A recent review paper describes many types of remote sensors that can be used for demining, and discuss some algorithms for data fusion and image processing (Robledo et al. 2009). They explain how reflective infrared, thermal infrared, and hyperspectral methods can detect anomalous signals emitted or reflected over the mine surface or soil immediately over the mine. Of course, data processing and fusion algorithms will finally determine whether an object's image corresponds to a landmine. Although these different spectral images can be fused at the pixel level, to determine whether the anomaly is a landmine, normally the features are fused, and further fusion at the decision-making level is needed.

In conclusion, it can be noted that remote sensing data fusion has become the main approach for detecting landmines and minefields, as the humanitarian mine action community has come to the opinion that no single-sensor system is capable of detecting mines at the required accuracy/reliability levels over a wide range of terrain conditions. Many national organizations working on humanitarian demining have researchers working on algorithm development of single sensors for feature detection/extraction. Thus, remote sensing image and data fusion for landmine detection tends to include three main fusion levels (hybrid fusion). First, at the raw data level of each individual sensor, the detection of landmines is located. Then, these are fused by any of a number of available algorithms to have a feature-based fusion product. These features are then fused, which is further processed by means of decision-based rules, to come up with a final decision on whether the objects are landmines or not. This approach helps to reduce the number of "false alarms."

The level at which the fusion is done depends on whether the process relies mainly on expert knowledge of the image analyst or on statistical modeling of experimental data. The main statistical decision-based fusion algorithms used, based on a detailed literature survey, include Bayes, Dempster–Shafer, fuzzy probabilities, and rules (Breuers et al. 1999). For real-world applications in the mine-affected countries, the approach of using visual analysis and fusion of the pixel-based and feature-based fusion, followed up by the experience of the remote sensing image analyst, have given best results to date (ITC 1999). In such a way, the final fused system will be more robust to a variety of environmental conditions. Thus, remote sensing image fusion can greatly assist in area reduction, thus speeding up the demining process and making land reusable again by the local population as soon as possible.

The variety of types of terrain in which landmines are found, the numerous different types of landmines, and the environmental/vegetation conditions all make it very difficult to detect buried or surface-laid mines. The numerous "false alarms" that most airborne systems produce make the job of the image analyst extremely difficult. No single mine sensor has the potential to increase the probability of detection and decrease false alarm rates for all types of mines under the wide variety of environmental conditions in which mines exist. Rather than focusing on individual technologies operating in isolation, the design of an integrated, multisensory system that would overcome the limitations of any single-sensor technology is the way that new developments in humanitarian demining are heading. Hence, for the landmine problem, it is mostly tackled by higher level feature level or decision level fusion rather than pixel-based fusion techniques, as often not only imaging sensors but also several other mine-detecting techniques that provide other nonimage signals, such as metal detectors and vapor sensors, are used.

6.12 Oceanography

Oceans cover more than 70% of Earth's surface, and hence have been an important application field for image fusion studies. It was one of the first application areas to use remote sensing image fusion to study topics such as ocean waves, ocean currents, and sea surface temperatures. The main satellites used in the late 1908s and early 1990s for image fusion studies were the SEASAT (L-Band), ALMAZ (S-Band), and the reflective infrared band of Landsat 4 (0.63–0.69 µm). As these had similar spatial resolutions, they could be fused for global surface ocean wave fields and for Polar sea ice studies (Tilley 1992; Tilley et al. 1992). By fusing different SAR systems with different wavelengths with the optical imagery from Landsat, it became possible to determine various geophysical ocean wave quantities such as significant wave height and mean square surface slope, items that were not possible to obtain from a single sensor alone. A very important application in the field of oceanography is the study of sea ice. This is of vital relevance to shipping, and especially, now that with climate change, the northern Arctic shipping routes are becoming more used. Hence, using mainly various combinations of SAR data (e.g., ERS and Radarsat), it is possible to map sea ice on a regular basis. Optical data in the Arctic regions are not possible for large parts of the year due to darkness and weather conditions. Especially, during the winter months, one can use a multi-sensor data fusion approach based on a multilayer neural network to classify sea ice. Using multiple sensors, a substantial improvement can be obtained, compared to single-sensor data, because of the complementarity of the different SAR parameters on the various operational SAR satellite systems (Bogdanov et al. 2005).

RSIF for oceanographic applications are very valuable as inputs to many oceanographic studies, including ocean wave forecasting models and for modeling surface currents, sea surface temperature, sea surface height determination, ocean tide monitoring, ocean circulation, sea surface salinity mapping, sea ice, chlorophyll concentrations, and many other parameters. For each of these parameters, different satellites and different sensors have been put in space to measure these aspects, such as soil moisture and ocean salinity (SMOS) for salinity, MODIS and AVHRR for temperature, TOPEX/ Poseidon, ERS-1, JASON-1 and 2, HY-2A, etc. for sea surface height, and other parameters. However, most sensors only measure one parameter (e.g., temperature or wave height with altimeters, or similar). Hence, data fusion has proven to be a most valuable approach to such integrated oceanographic studies (Raizer 2013; Teggi and Despini 2014; Umbert et al. 2014, 2015; Fok 2015; Kim et al. 2015).

6.13 Security and Defense

In the context of national security and military applications, multi-sensor remote sensing plays an important role. The data are part of the geospatial and imagery intelligence used for crisis management missions and operations as well as maritime surveillance and border control. The processing techniques serve automatic target recognition and mapping of areas and developments of interest.

A simple RGB overlay served a complex urban and harbor area (Oslo, Sweden) application using multi-temporal and multi-sensor SAR. The solution enabled shortening the repeat cycle of data acquisition because two different radar satellites delivered the data (Kempf and Anglberger 2013).

6.14 Missing Information Reconstruction

The problem of missing data within an image of a certain area is manifold. The origin of such noninformation pixels in VIR data is related to sensor failure or noise on one hand, or cloud cover, cloud shadows, and other atmospheric conditions on the other hand if the study of interest is not related to meteorology, atmosphere, or water. Apparently, about 35% of all acquisitions over land are cloud covered (Ju and Roy 2008). In the humid tropics, the percentage is higher. For Canada, studies revealed more than 50% (Cihlar 1987). In addition to the cloud cover problem, malfunctioning of the sensor is a major issue. In the case of SAR, the distortions (foreshortening, shadow, layover) can be so heavy over high terrain height variations that the information is useless.

Reconstruction of missing information can be categorized into four groups (Shen et al. 2015):

1. Spatial-based methods (without other information)
2. Spectral-based methods (complementary information from other spectra)
3. Temporal-based methods (multi-temporal acquisitions)
4. Hybrid methods (combinations of the above)

One method of overcoming the common problem of cloud removal in a multi-temporal data set is to use an image fusion methodology. Cloud removal is of major interest in optical remote sensing, and many different methods have been suggested, for example, thresholding, wavelet decomposition, and others. Problems occur with radiance variations if radiance is the major parameter in cloud detection. Cloud shadows provide additional challenges. There are single- and multi-sensor solutions. A single-sensor, intensity-insensible method consists of applying a 1D pseudo-Wigner distribution (PWD) transformation on all the source images, and using a pixel-wise cloud model. This is a type of "de-noising" method that aids in producing a 2D clean image (Gabarda and Cristóbal 2007). Other research results suggest multi-sensor RSIF, taking advantage of cloud-cover independent radar images to replace missing information in optical imagery. In a case study, the optical image (SPOT XS) density slicing including all three multispectral bands forms a cloud mask. Some postprocessing improves the cloud mask applying a clump filter to eliminate small patches that are misclassified as "cloud." Later on, a region-growing process ensures that the very borders of the clouds are included in the mask since they contain mixed pixels that falsify the original spectral information of the land cover pixels (Pohl 1996). Figure 6.14 illustrates the cloud masking result on a SPOT scene covering an area on Sumatra, Indonesia.

(a)   (b)

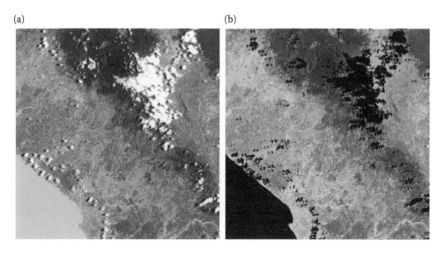

FIGURE 6.14
Overcoming cloud cover using multi-sensor RSIF: (a) original SPOT XS, (b) cloud masked SPOT XS (Pohl 1996). (Images courtesy of SPOT Image.)

The no-value areas, previously covered by clouds, are then filled with SAR imagery, in this case using ERS-1 SAR data. Similarly, JERS-1 SAR data were inserted to replace foreshortening and layover areas in the ERS-1 data. The final fused product based on a hybrid RSIF approach involving mosaicking and multiplication takes advantage of VIR/SAR fusion overcoming lack of data through clouds as depicted in Figure 6.15a and b (Pohl 1996).

6.15 Other Applications

There are numerous other applications of RSIF. Almost every application of remote sensing will benefit from a fusion approach, in that looking at a particular object or topic with multi-spatial, multispectral, multi-temporal image fusion approach will, in almost all cases, provide more information about the subject of interest than is obtainable from a single image obtained from a single sensor at only one point in time.

One interesting application is that of archaeology. For archaeology, similar approaches to that described above in the sections on underground coal fires and landmine detection are used. That is so because any object buried beneath the surface implies that the vegetation and or soil above the object of interest have been disturbed, resulting in changes in the vegetation, soil structure, soil moisture, surface temperature, etc. Hence, for example, by fusing data from reflective infrared for changes in vegetation, with thermal infrared for changes in temperature, and with active microwaves such as

(a)   (b)

(c)   (d)

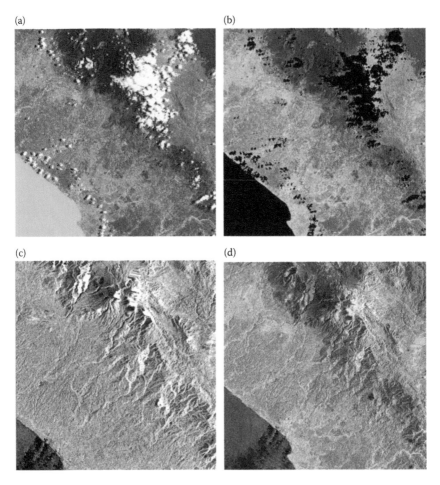

FIGURE 6.15
(a) Original SPOT XS image; (b) cloud removal in SPOT XS; (c) replacement image consisting of fused ERS-1/JERS-1 SAR; (d) cloud-free, fused multi-sensor image of SPOT XS, ERS-1 SAR, and JERS-1 SAR (Pohl 1996). (SPOT image courtesy of SPOT Image; ERS image courtesy of ESA; JERS-1 image courtesy of NASDA [now: JAXA].)

ground-penetrating radar (GPR), one can detect and identify many more features of such archaeological sites than from any single sensor. Archaeologists using remote sensing techniques to discover new sites or to monitor existing known sites use the typical image characteristics of tone, texture, structure, pattern, shape, size, shadow, orientation, relief, lineation, crop/soil marks, and associated features from the fused images for their research. The platforms commonly used range from on-site LiDAR and cameras, through low-altitude-tethered balloons for detailed site studies, through the use of various airborne sensors, to Earth observation satellites for regional studies (Karamitrou et al. 2011; Yu et al. 2012; Verhoeven 2015).

Other applications that have successfully used RSIF to solve problems are ground deformation mapping (Zhang et al. 2015), lake monitoring (Muad and Foody 2012), vehicle control on highways (Hofmann et al. 2001), and pollution studies (Zia et al. 2002).

6.16 Summary

From the description of the different use of RSIF, it is obvious that there is no generic fusion system. The applications provide the key parameters in the decision on how to process the data to extract the needed information. Therefore, an understanding of the relevant variables in an application context is crucial. We have compiled a table listing the most suitable data collection along with the parameters relevant to the various applications presented in this chapter (see Table 6.3). Along with it, successful fusion requires complex processing. With more sophisticated techniques and the ability to process large volumes of data (see Chapter 7 for details), researchers are moving toward the hybrid approach to fulfill application requirements. Fusion processing takes place at all three levels (image, feature, and information levels), and very diverse data are being integrated. A very good example is a recent study in a coastal zone, which is very complex by nature. In order to map benthic habitat and monitor coral reefs, hyperspectral images, aerial photography, and bathymetric data were fused. While image and data fusion, OBIA, and machine learning algorithms have been widely used in terrestrial remote sensing, as described in Sections 6.3 and 6.6, they have not yet been widely used for marine environments. Zhang (2015) developed a methodology to do pixel- and feature-level fusion by merging high-resolution (1 m) aerial photographs with 17 m spatial resolution AVIRIS hyperspectral imagery, along with 10 m spatial resolution bathymetric data. He showed that such an image and data fusion approach can make a good contribution to benthic habitat classification in a coral reef ecosystem. Pixel level fusion supported the increase in spatial resolution. Feature level fusion identifies relevant features for further evaluation. Decision level fusion is then applied to produce the final maps and uniquely classify the data to provide the necessary information. This sophisticated fusion scheme is able to overcome the limitations of the individual approach and benefits from the advantages of each individual input data. It proves that the better we can describe the real world, the better we will understand the complex processes. Hence, it is important to have a deep understanding of the problem being investigated. Once that is clear, one can select the optimum data sources required to solve the problem, and decide on the best fusion approach to extract the maximum useful information from the various input data sources.

TABLE 6.3

Data Selection and Parameters for Different Applications of RSIF

Data	Parameters
1. Urban Studies	
Multi-temporal SAR, MODIS, and DMSP	Density of built-up area
High-resolution optical and LiDAR	Structural data and height of buildings
QuickBird and TerraSAR-X	Urban land cover classification
SPOT and Radarsat-2	Temporal urban land cover changes
Worldview multi-angle imagery	Urban features such as skyscrapers, bridges, high-volume highways, and car parks
2. Agriculture	
Very high-resolution data	Precision agriculture
Multi-temporal VIR	Phenology of planted crops
VIR/SAR	Changes in cropland extent
Pansharpened Landsat	Sugarcane mapping
MODIS TIR and Landsat	Evapotranspiration
UAV Hyperspectral and Formosat-2	Spectro-temporal reflectance surface
SPOT-5 and HJ-1A/B	Estimating wheat yield and protein content
3. LULC Mapping	
AVHRR, Landsat, SPOT VEGETATION, MODIS, MERIS, GLI	Long-term global land use and land cover changes
VIR/SAR	Improving land use classification results
ERS-1, JERS-1 with SPOT-XS	Map updating in tropical areas
Radarsat, PALSAR, and Landsat TM	Land cover classification
TerraSAR-X and Landsat TM	Protected area mapping
Hyperspectral data and LiDAR	Land cover classification
4. Geology	
SPOT, Landsat, and SEASAT	Geological structures
VIR/SAR	Metallic deposits in the tropics
Multi-temporal VIR	Subsurface structural mapping
Band ratioing	Lithological units mapping
ASTER fusion of VNIR, SWIR, and TIR	Enhanced mineral mapping
Hyperspectral	Rock and soil chemistry
Landsat TM, ASTER, SRTM, and airborne magnetic and gamma-ray spectrometry	Gold mineralization and geological map updating
SAR and gamma-ray spectrometry	Topographic, vegetation, and lithological information
Landsat ETM and ERS-2 SAR	Detection of bauxite mineralized zones
Landsat ETM and Radarsat-1	Morphotectonics and spectral properties of gold
5. Vegetation Mapping	
High spatial and high temporal fusion	Carbon budget calculations, soil monitoring, land use/land cover changes
Fusion of vegetation indices	Photosynthesis, plant growth

(Continued)

TABLE 6.3 (*Continued*)

Data Selection and Parameters for Different Applications of RSIF

Data	Parameters
Landsat and MODIS	Vegetation phenology
RapidEye and MODIS	Vegetation dynamics
IKONOS and Landsat	Forest texture and vegetation phenology
VIR/SAR	Class discrimination and feature enhancement
MODIS and Radarsat-2	Grassland alfalfa
Landsat and ALOS PALSAR and Radarsat-2	Vegetation classification in tropics
UAV imagery with TerraSAR-X	Vegetation structure in wetlands
6. Forest Monitoring	
Pansharpening (single sensor)	Higher classification accuracy
VIR/SAR	Discrimination of different tree types
ALOS PALSAR, LiDAR, and hyperspectral data	Forest classification, forest species, land cover, and vertical forest structures
SPOT, ERS, and Radarsat-1	Complementing missing information in tropical forests, rubber plantation mapping
Multi-temporal SAR data	Land cover mapping in tropics
Landsat and ALOS PALSAR	Deforestation and oil palm mapping
Landsat TM and ENVISAT	Forest taxonomy
7. Natural Hazards and Disasters	
VIR/SAR	Flood monitoring
Landsat and ERS-1	Flood impact and extent
Multi-temporal SAR	Volcanic ash clouds, gas emissions
ALOS-2, Radarsat-2, and TerraSAR-X	Volcano monitoring
AVHRR, ERS-2, Radarsat-1, and ALOS PALSAR	Mapping volcanic lahar and lava flows
Landsat and SPOT	Landslide detection and monitoring
NDVI, DTM, and optical data	Landslide probability
8. Coastal Zones	
Multi-temporal Landsat	Coastal changes
Multi-temporal IKONOS	Shoreline change monitoring
Multi-temporal SAR	Shoreline mapping in high latitudes
LiDAR and historical aerial photography	Shoreline changes
IKONOS and LiDAR	Coastal mapping
AVIRIS, LiDAR, and DEMs	Detailed coastal mapping
RaidEye and TerraSAR-X	Coastal vegetation, shellfish beds, sediments
SPOT, Landsat ETM + , LiDAR, aerial photography, and maps	Wetland distribution and coastal topography
9. Coal Fires	
Airborne and satellite optical, thermal data sets	Detection, measurement, and monitoring of surface, subsurface fires

(*Continued*)

TABLE 6.3 (*Continued*)

Data Selection and Parameters for Different Applications of RSIF

Data	Parameters
Optical, thermal night time imagery plus SAR	Land subsidence, fire hot spots
AVHRR, SPOT VEGETATION, Landsat TIR band, ASTER TIR bands	Size, depth of fires, and direction of burning
Landsat TM day time, Landsat TM night time TIR, DTM	Relation among mining areas, coal fires, and land subsidence
Fusion of coal fire-risk areas with thermal anomaly areas	Automatic detection of coal fires
10. Landmine Detection	
Multi-temporal, multiscale, multispectral data sets	Detection of minefields by fusing imagery before, during, and after minefield laying
Landsat TM, historical aerial photography, and KFA imagery	Border minefields
Multi-platform fusion (vehicle, UAV, helicopter, aircraft, satellite)	Detection of anomalous signals of landmines
11. Oceanography	
SEASAT, ALMAZ, and Landsat	Ocean waves, ocean currents, polar sea ice
MODIS and AVHRR	Sea surface temperature
TOPEX/Poseidon and ERS-1 and JASON-1	Sea surface height

References

Abdelkareem, M. and F. El-Baz. 2014. Analyses of optical images and radar data reveal structural features and predict groundwater accumulations in the central Eastern Desert of Egypt. *Arabian Journal of Geosciences* 8 (11):2653–2666.

Alexander, D. E. 1993. *Natural Disasters*. Berlin, Germany: Springer Science & Business Media.

Alonzo, M., B. Bookhagen, and D. A. Roberts. 2014. Urban tree species mapping using hyperspectral and LiDAR data fusion. *Remote Sensing of Environment* 148:70–83.

Amarsaikhan, D., T. Bat-Erdene, M. Ganzorig, and B. Nergui. 2013. Knowledge-based classfication of Quickbird image of Ulaanbaatar city, Mongolia. *American Journal of Signal Processing* 3 (3):71–77.

Amarsaikhan, D., V. Battsengel, G. Bolor, D. Enkhjargal, and E. Jargaldadai. 2015. Fusion of optical and SAR images for enhancement of forest classes. In *36th Asian Conference on Remote Sensing (ACRS) 'Fostering Resilient Growth in Asia'*. Manila, Philippines: AARS.

Amarsaikhan, D., H. H. Blotevogel, J. L. van Genderen, M. Ganzorig, R. Gantuya, and B. Nergui. 2010. Fusing high-resolution SAR and optical imagery for improved urban land cover study and classification. *International Journal of Image and Data Fusion* 1 (1):83–97.

Amarsaikhan, D. and N. Ganchuluun. 2015. Fusion and classification of multisource images for update of forest GIS. In *Image Fusion: Principles, Technology and Applications*, edited by C. T. Davis. New York, USA: Nova Science Publisher, pp. 83–121.

Amarsaikhan, D., J. Janzen, E. Egshiglen, and R. Gantuya. 2014. Urban land use change study in Mongolia using spatial techniques. *International Journal of Sustainable Building Technology and Urban Development* 5 (1):35–43.

Amorós-López, J., L. Gómez-Chova, L. Alonso et al. 2013. Multitemporal fusion of Landsat/TM and ENVISAT/MERIS for crop monitoring. *International Journal of Applied Earth Observation and Geoinformation* 23 (1):132–141.

Amoros-Lopez, J., L. Gomez-Chova, L. Alonso, L. Guanter, J. Moreno, and G. Camps-Valls. 2011. Regularized multiresolution spatial unmixing for ENVISAT/MERIS and Landsat/TM image fusion. *IEEE Geoscience and Remote Sensing Letters* 8 (5):844–848.

Angel, S., J. Parent, D. L. Civco, and A. M. Blei. 2012. *Atlas of Urban Expansion.* Cambridge, MA, USA: Lincoln Institute of Land Policy.

Arenas-Castro, S., Y. Julien, J. C. Jiménez-Muñoz, J. A. Sobrino, J. Fernández-Haeger, and D. Jordano-Barbudo. 2013. Mapping wild pear trees (*Pyrus bourgaeana*) in Mediterranean forest using high-resolution QuickBird satellite imagery. *International Journal of Remote Sensing* 34 (9–10):3376–3396.

Aydal, D., E. Arda, and Ö. Dumanlilar. 2007. Application of the Crosta technique for alteration mapping of granitoidic rocks using ETM+ data: Case study from eastern Tauride belt (SE Turkey). *International Journal of Remote Sensing* 28 (17):3895–3913.

Bahiru, E. A. and T. Woldai. 2016. Integrated geological mapping approach and gold mineralization in Buhweju area, Uganda. *Ore Geology Reviews* 72 (Part 1):777–793.

Balik Sanli, F., Y. Kurucu, and M. Esetlili. 2009. Determining land use changes by radar-optic fused images and monitoring its environmental impacts in Edremit region of western Turkey. *Environmental Monitoring and Assessment* 151 (1–4):45–58.

Basuki, T. M., A. K. Skidmore, Y. A. Hussin, and I. van Duren. 2013. Estimating tropical forest biomass more accurately by integrating ALOS PALSAR and Landsat-7 ETM+ data. *International Journal of Remote Sensing* 34 (13):4871–4888.

Berger, C., M. Voltersen, R. Eckardt et al. 2013a. Multi-modal and multi-temporal data fusion: Outcome of the 2012 GRSS data fusion contest. *IEEE Journal of Selected Topics in Applied Earth Observations and Remote Sensing* 6 (3):1324–1340.

Berger, C., M. Voltersen, S. Hese, I. Walde, and C. Schmullius. 2013b. Robust extraction of urban land cover information from HSR multi-spectral and LiDAR data. *IEEE Journal of Selected Topics in Applied Earth Observations and Remote Sensing* 6 (5):2196–2211.

Bhandari, S., S. Phinn, and T. Gill. 2012. Preparing Landsat image time series (LITS) for monitoring changes in vegetation phenology in Queensland, Australia. *Remote Sensing* 4 (6):1856–1886.

Bloch, I., N. Milisavljevic, and M. Acheroy. 2007. Multisensor data fusion for space-borne and airborne reduction of mine suspected areas. *International Journal of Advanced Robotics Systems* 4 (2):173–186.

Bogdanov, A. V., S. Sandven, O. M. Johannessen, V. Y. Alexandrov, and L. P. Bobylev. 2005. Multisensor approach to automated classification of sea ice image data. *IEEE Transactions on Geoscience and Remote Sensing* 43 (7):1648–1664.

Bovolo, F., L. Bruzzone, L. Capobianco, A. Garzelli, S. Marchesi, and F. Nencini. 2010. Analysis of the effects of pansharpening in change detection on VHR images. *IEEE Geoscience and Remote Sensing Letters* 7 (1):53–57.

Breuers, M. G. J., P. B. W. Schwering, and S. P. van den Broek. 1999. Sensor fusion algorithms for the detection of land mines. Paper read at *Detection and Remediation Technologies for Mines and Minelike Targets IV*, 2 August 1999, Orlando, Florida, USA.

Bruzzone, L. and D. F. Prieto. 2000. Automatic analysis of the difference image for unsupervised change detection. *IEEE Transactions on Geoscience and Remote Sensing* 38 (3):1171–1182.

Busetto, L., M. Meroni, and R. Colombo. 2008. Combining medium and coarse spatial resolution satellite data to improve the estimation of sub-pixel NDVI time series. *Remote Sensing of Environment* 112 (1):118–131.

Cammalleri, C., M. C. Anderson, F. Gao, C. R. Hain, and W. P. Kustas. 2014. Mapping daily evapotranspiration at field scales over rainfed and irrigated agricultural areas using remote sensing data fusion. *Agricultural and Forest Meteorology* 186:1–11.

Castañeda, C. and D. Ducrot. 2009. Land cover mapping of wetland areas in an agricultural landscape using SAR and Landsat imagery. *Journal of Environmental Management* 90 (7):2270–2277.

Celik, T. and M. Kai-Kuang. 2011. Multitemporal image change detection using undecimated discrete wavelet transform and active contours. *IEEE Transactions on Geoscience and Remote Sensing* 49 (2):706–716.

Chang, L.-Y., A. Chen, C. Chen, and C. Huang. 1999. A robust system for shoreline detection and its application to coastal-zone monitoring. In *20th Asian Conference on Remote Sensing (ACRS)*. Hong Kong, China: AARS.

Chang, Y., Y. Wang, Y. Fu, C. Han, J. Chanussot, and B. Huang. 2014. Multisource data fusion and fisher criterion-based nearest feature space approach to landslide classification. *IEEE Journal of Selected Topics in Applied Earth Observations and Remote Sensing* 8 (2):576–588.

Chavez, P. 1986. Digital merging of Landsat TM and digitized NHAP data for 1: 24,000-scale image mapping (National High Altitude Program). *Photogrammetric Engineering and Remote Sensing* 52 (10):1637–1646.

Chavez, P., S. C. Sides, and J. A. Anderson. 1991. Comparison of three different methods to merge multiresolution and multispectral data-Landsat TM and SPOT panchromatic. *Photogrammetric Engineering and Remote Sensing* 57 (3):295–303.

Chavez Jr, P. S. and A. Y. Kwarteng. 1989. Extracting spectral contrast in Landsat Thematic Mapper image data using selective principal component analysis. *Photogrammetric Engineering and Remote Sensing* 55 (3):339–348.

Chen, J. M. and T. A. Black. 1992. Defining leaf area index for non-flat leaves. *Plant, Cell & Environment* 15 (4):1365–3040.

Chen, L. C. 1998. Detection of shoreline changes for tideland areas using multi-temporal satellite images. *International Journal of Remote Sensing* 19 (17):3383–3397.

Cihlar, J. 1987. Remote sensing of global change: An opportunity for Canada. *Proceedings of the 11th Canadian Symposium on Remote Sensing*, Waterloo, Ontario, Canada, pp. 39–48.

Cihlar, J. 2000. Land cover mapping of large areas from satellites: Status and research priorities. *International Journal of Remote Sensing* 21 (6–7):1093–1114.

Cook, G. 2003. Evaluating LiDAR for documenting shoreline change. In *3rd Biennial Coastal GeoTools Conference*. Charleston, SC.

Coppin, P., I. Jonckheere, K. Nackaerts, B. Muys, and E. Lambin. 2004. Digital change detection methods in ecosystem monitoring: A review. *International Journal of Remote Sensing* 25 (9):1565–1596.

Crosta, A. and J. M. C. M. Moore. 1989. Enhancement of Landsat Thematic Mapper imagery for residual soil mapping in SW Minas Gerais State, Brazil—A prospecting case history in greenstone belt terrain. Paper read at *7th Thematic Conference on Remote Sensing for Exploration Geology—Methods, Integration, Solutions,* 2–6 October 1989, Calgary, Alberta, Canada.

Cudahy, T. and R. Hewson. 2002. ASTER geological case histories: Porphyry-skarnepithermal, iron oxide Cu-Au and Broken hill Pb-Zn-Ag. Paper read at *Annual General Meeting of the Geological Remote Sensing Group 'ASTER Unveiled',* Burlington House, Piccadilly, London, UK.

Dahiya, S., P. K. Garg, and M. K. Jat. 2013. A comparative study of various pixel-based image fusion techniques as applied to an urban environment. *International Journal of Image and Data Fusion* 4 (3):197–213.

de Almeida Furtado, L. F., T. S. Freire Silva, P. J. Farias Fernandes, and E. M. Ledo de Moraes Novo. 2015. Land cover classification of Lago Grande de Curuai floodplain (Amazon, Brazil) using multi-sensor and image fusion techniques. *Acta Amazonica* 45 (2):195–202.

Dean, K. G., J. Dehn, K. Engle, P. Izbekov, K. Papp, and M. Patrick. 2002. Operational satellite monitoring of volcanoes at the Alaska Volcano Observatory. *Advances in Environmental Monitoring and Modelling* 1 (3):70–97.

de Beurs, K. M. and G. M. Henebry. 2004. Land surface phenology, climatic variation, and institutional change: Analyzing agricultural land cover change in Kazakhstan. *Remote Sensing of Environment* 89 (4):497–509.

Demargne, L., E. Nezry, F. Yakam-Simen, and E. Nabet. 2001. Use of SPOT and radar data for forest inventory in Sarawak, Malaysia. In *Paper presented at the 22nd Asian Conference on Remote Sensing, November 5–9, 2001, Singapore. Centre for Remote Imaging, Sensing and Processing (CRISP), National University of Singapore; Singapore Institute of Surveyors and Valuers (SISV); Asian Association on Remote Sensing (AARS),* Singapore.

Demographia. 2015. *Demographia World Urban Areas (Built-Up Urban Areas or World Agglomerations).* Belleville, IL: Wendell Cox. www.demographia.com.

Deng, J. S., K. Wang, Y. H. Deng, and G. J. Qi. 2008. PCA-based land-use change detection and analysis using multitemporal and multisensor satellite data. *International Journal of Remote Sensing* 29 (16):4823–4838.

Doña, C., N.-B. Chang, V. Caselles et al. 2015. Integrated satellite data fusion and mining for monitoring lake water quality status of the Albufera de Valencia in Spain. *Journal of Environmental Management* 151:416–426.

Dong, J., X. Xiao, B. Chen et al. 2013. Mapping deciduous rubber plantations through integration of PALSAR and multi-temporal Landsat imagery. *Remote Sensing of Environment* 134:392–402.

Doraiswamy, P. C., J. L. Hatfield, T. J. Jackson, B. Akhmedov, J. Prueger, and A. Stern. 2004. Crop condition and yield simulations using Landsat and MODIS. *Remote Sensing of Environment* 92 (4):548–559.

Drzyzga, S. and A. Arbogast. 2002. Technical notes about using IKONOS imagery to monitor and map coastal landscape change along the great lakes. In *36th North-Central Section and the 51st South-Eastern Section GSA Joint Annual Meeting.* Lexington, Kentucky.

Elaksher, A. F. 2008. Fusion of hyperspectral images and LiDAR-based dems for coastal mapping. *Optics and Lasers in Engineering* 46 (7):493–498.

Erasmi, S. and A. Twele. 2009. Regional land cover mapping in the humid tropics using combined optical and SAR satellite data—A case study from Central Sulawesi, Indonesia. *International Journal of Remote Sensing* 30 (10):2465–2478.

Ferro, A., D. Brunner, and L. Bruzzone. 2009. An advanced technique for building detection in VHR SAR images. Paper read at *Image and Signal Processing for Remote Sensing XV*, 31 August 2009, Berlin, Germany.

Fok, H. S. 2015. Data fusion of multisatellite altimetry for ocean tides modelling: A spatio-temporal approach with potential oceanographic applications. *International Journal of Image and Data Fusion* 6 (3):232–248.

Friedl, M. A., D. Sulla-Menashe, B. Tan et al. 2010. MODIS Collection 5 global land cover: Algorithm refinements and characterization of new datasets. *Remote Sensing of Environment* 114 (1):168–182.

Furtado, L. F. d. A., T. S. F. Silva, P. J. F. Fernandes, and E. M. L. d. M. Novo. 2015. Land cover classification of Lago Grande de Curuai floodplain (Amazon, Brazil) using multi-sensor and image fusion techniques. *Acta Amazonica* 45 (2):195–202.

Gabarda, S. and G. Cristóbal. 2007. Cloud covering denoising through image fusion. *Image and Vision Computing* 25 (5):523–530.

Gamba, P. 2014. Image and data fusion in remote sensing of urban areas: Status issues and research trends. *International Journal of Image and Data Fusion* 5 (1):2–12.

Gao, F., T. Hilker, X. Zhu et al. 2015. Fusing Landsat and MODIS data for vegetation monitoring. *IEEE Geoscience and Remote Sensing Magazine* 3 (3):47–60.

Genderen, J. L. van. 2004. An integrated global observing strategy for earthquake prediction. In *IGOS International Workshop on 'Towards the Implementation of an Integrated Global Observing Strategy*. Tokyo, Japan.

Genderen, J. L. van. 2005. The use of a data fusion approach to integrate multiple earthquake precursors into a robust prediction methodology. In *First DEMETER Investigators Workshop*, edited by CNES. Paris, France: CNES.

Gens, R. *Interactive World Map of Coal and Peat Fires*. Elsevier 2014 [cited 10 November 2015]. Available from http://booksite.elsevier.com/9780444595096/interactive_map.php

Gevaert, C. M., J. Suomalainen, T. Jing, and L. Kooistra. 2015. Generation of spectral-temporal response surfaces by combining multispectral satellite and hyper-spectral UAV imagery for precision agriculture applications. *IEEE Journal of Selected Topics in Applied Earth Observations and Remote Sensing* 8 (6):3140–3146.

Ghanbari, Z. and M. Sahebi. 2014. Improved IHS algorithm for fusing high resolution satellite images of urban areas. *Journal of the Indian Society of Remote Sensing* 42 (4):689–699.

Gibeaut, J. C., W. A. White, T. Hepner et al. 2000. Texas shoreline change project; Gulf of Mexico shoreline change from the Brazos River to Pass Cavallo. Texas Coastal Coordination Council.

Gill, J. C. and B. D. Malamud. 2014. Reviewing and visualizing the interactions of natural hazards. *Reviews of Geophysics* 52 (4):680–722.

Goetz, A. F. H. and L. C. Rowan. 1981. Geologic remote sensing. *Science* 211 (4484):781–791.

Goodenough, D. G., C. Hao, A. Dyk, A. Richardson, and G. Hobart. 2008. Data fusion study between polarimetric SAR, hyperspectral and LiDAR data for forest information. Paper read at *IEEE International Geoscience and Remote Sensing*

Symposium, 2008. IGARSS 2008 7–11 July 2008, Boston, Massachusets, USA, 2:II-281–II-284.

Gozzard, J. R. 2004. Part 2 Predictive Regolith-landform mapping. In *SEG Workshop—How to Look at, in and through the Regolith for Effective Predictive Mineral Discoveries.* Perth, Western Australia.

Gray, J. and C. Song. 2012. Mapping leaf area index using spatial, spectral, and temporal information from multiple sensors. *Remote Sensing of Environment* 119:173–183.

Green, D. 2015. Viewing the earth's global environment from space: From scientific knowledge to social benefits. In *9th International Symposium on Digital Earth,* Hugh Millward, Dirk Werle and Danika van Proosdij (Eds). Halifax, Canada: IOP Conference Series: Earth and Environmental Science.

Gungor, O. and O. Akar. 2010. Multi sensor data fusion for change detection. *Scientific Research and Essays* 5 (18):2823–2831.

Hankui, Z., J. M. Chen, H. Bo, S. Huihui, and L. Yiran. 2014. Reconstructing seasonal variation of Landsat vegetation index related to leaf area index by fusing with MODIS data. *IEEE Journal of Selected Topics in Applied Earth Observations and Remote Sensing* 7 (3):950–960.

Hansen, M. C., P. V. Potapov, R. Moore et al. 2013. High-resolution global maps of 21st-century forest cover change. *Science* 342 (6160):850–853.

Harbi, H. and A. Madani. 2013. Utilization of SPOT 5 data for mapping gold mineralized diorite–tonalite intrusion, Bulghah gold mine area, Saudi Arabia. *Arabian Journal of Geosciences* 7(9):3829–3838.

Henderson, F. M., T. F. Hart Jr, R. Chasan, and J. Portolese. 1998. Analysis of optical-SAR data fusion and merged data set procedures in highly developed urbanized coastal ecosystems. In *International Geoscience and Remote Sensing Symposium (IGARSS).* Seattle, WA: IEEE.

Hill, D. 2005. *Handbook of Biodiversity Methods: Survey, Evaluation and Monitoring.* Cambridge, England: Cambridge University Press.

Hofmann, U., A. Rieder, and E. D. Dickmanns. 2001. Radar and vision data fusion for hybrid adaptive cruise control on highways. In *Computer Vision Systems,* edited by B. Schiele and G. Sagerer. Berlin, Heidelberg: Springer.

Holmström, H. and J. E. S. Fransson. 2003. Combining remotely sensed optical and radar data in kNN-estimation of forest variables. *Forest Science* 49 (3):409–418.

Hong, G., A. Zhang, F. Zhou, and B. Brisco. 2014. Integration of optical and synthetic aperture radar (SAR) images to differentiate grassland and alfalfa in Prairie area. *International Journal of Applied Earth Observation and Geoinformation* 28 (1):12–19.

Hu, Y., G. Jia, C. Pohl et al. 2015. Improved monitoring of urbanization processes in China for regional climate impact assessment. *Environmental Earth Sciences* 73 (12):8387–8404.

ITC. 1999. Final Report Airborne Minefield Detection. Enschede, The Netherlands: International Institute for Aerospace Survey and Earth Sciences (ITC).

Jawak, S. D. and A. J. Luis. 2013. A spectral index ratio-based Antarctic land-cover mapping using hyperspatial 8-band WorldView-2 imagery. *Polar Science* 7 (1):18–38.

Jensen, J. R. and J. Im. 2007. Remote sensing change detection in urban environments. In *Geo-Spatial Technologies in Urban Environments,* edited by R. R. Jensen, J. D. Gatrell, and D. McLean. Berlin, Heidelberg: Springer, pp. 7–31.

Ji, M. and J. R. Jensen. 1999. Effectiveness of subpixel analysis in detecting and quantifying urban imperviousness from Landsat thematic mapper imagery. *Geocarto International* 14 (4):33–41.

Johnson, B. A., H. Scheyvens, and B. R. Shivakoti. 2014. An ensemble pansharpening approach for finer-scale mapping of sugarcane with Landsat 8 imagery. *International Journal of Applied Earth Observation and Geoinformation* 33:218–225.

Ju, J. and D. P. Roy. 2008. The availability of cloud-free Landsat ETM+ data over the conterminous United States and globally. *Remote Sensing of Environment* 112 (3):1196–1211.

Jung, R., W. Adolph, M. Ehlers, and H. Farke. 2015. A multi-sensor approach for detecting the different land covers of tidal flats in the German Wadden Sea—A case study at Norderney. *Remote Sensing of Environment* 170:188–202.

Karamitrou, A., M. Petrou, and G. Tsokas. 2011. Fusion of geophysical images in the study of archaeological sites. Paper read at *AGU Fall Meeting Abstracts*, San Francisco, California, USA.

Kavak, K. 2005. Determination of palaeotectonic and neotectonic features around the Menderes Massif and the Gediz Graben (western Turkey) using Landsat TM image. *International Journal of Remote Sensing* 26 (1):59–78.

Kempf, T. and H. Anglberger. 2013. Image fusion of different spaceborne SAR sensors for change detection. Paper read at *IEEE Radar Conference (RADAR), 2013* April 29 2013–May 3 2013.

Khandelwal, P., K. K. Singh, B. K. Singh, and A. Mehrotra. 2013. Unsupervised change detection of multispectral images using wavelet fusion and kohonen clustering network. *International Journal of Engineering and Technology* 5 (2):1401–1406.

Kim, M., J. Im, H. Han et al. 2015. Landfast sea ice monitoring using multisensor fusion in the Antarctic. *Giscience & Remote Sensing* 52 (2):239–256.

Komp, K.-U. 2015. High resolution land cover/land use mapping of large areas–current status and upcoming trends. *Photogrammetrie-Fernerkundung-Geoinformation* 2015 (5):395–410.

Kuenzer, C., J. Zhang, J. Li, S. Voigt, H. Mehl, and W. Wagner. 2007. Detecting unknown coal fires: Synergy of automated coal fire risk area delineation and improved thermal anomaly extraction. *International Journal of Remote Sensing* 28 (20):4561–4585.

Kuria, D. N., G. Menz, S. Misana et al. 2014. Seasonal vegetation changes in the Malinda wetland using bi-temporal, multi-sensor, very high resolution remote sensing data sets. *Advances in Remote Sensing* 3 (1):33–48.

Langford, R. L. 2015. Temporal merging of remote sensing data to enhance spectral regolith, lithological and alteration patterns for regional mineral exploration. *Ore Geology Reviews* 68:14–29.

Lee, D. S. and J. Shan. 2003. Combining LiDAR elevation data and IKONOS multispectral imagery for coastal classification mapping. *Marine Geodesy* 26 (1–2):117–127.

Li, J. and W. Chen. 2005. A rule-based method for mapping Canada's wetlands using optical, radar and DEM data. *International Journal of Remote Sensing* 26 (22):5051–5069.

Li, X. and L. Wang. 2015. On the study of fusion techniques for bad geological remote sensing image. *Journal of Ambient Intelligence and Humanized Computing* 6 (1):141–149.

Longbotham, N., C. Chaapel, L. Bleiler, C. Padwick, W. J. Emery, and F. Pacifici. 2012. Very high resolution multiangle urban classification analysis. *IEEE Transactions on Geoscience and Remote Sensing* 50 (4):1155–1170.

Lu, D., M. Batistella, and E. Moran. 2008. Integration of Landsat TM and SPOT HRG images for vegetation change detection in the Brazilian Amazon. *Photogrammetric Engineering and Remote Sensing* 74 (4):421.

Lu, D., G. Li, and E. Moran. 2014a. Current situation and needs of change detection techniques. *International Journal of Image and Data Fusion* 5 (1):13–38.

Lu, D., G. Li, E. Moran, L. Dutra, and M. Batistella. 2011. A comparison of multisensor integration methods for land cover classification in the Brazilian Amazon. *GIScience & Remote Sensing* 48 (3):345–370.

Lu, D., G. Li, E. Moran, and W. Kuang. 2014b. A comparative analysis of approaches for successional vegetation classification in the Brazilian Amazon. *GIScience & Remote Sensing* 51 (6):695–709.

Lu, D., P. Mausel, E. Brondízio, and E. Moran. 2004. Change detection techniques. *International Journal of Remote Sensing* 25 (12):2365–2401.

Maathuis, B. H. P. 2003. Remote sensing based detection of minefields. *Geocarto International* 18 (1):51–60.

Maathuis, B. H. P. 2008. *Remote Sensing for Area Reduction of Minefield Suspect Areas.* Saarbrücken, Germany: VDM Verlag Dr. Müller.

Maathuis, B. H. P. and J. L. van Genderen. 2004a. Remote sensing based detection of defensive minefields. *International Journal of Remote Sensing* 25 (23):5195–5200.

Maathuis, B. H. P. and J. L. van Genderen. 2004b. A review of satellite and airborne sensors for remote sensing based detection of minefields and landmines. *International Journal of Remote Sensing* 25 (23):5201–5245.

Maglione, P., C. Parente, and A. Vallario. 2015. High resolution satellite images to reconstruct recent evolution of domitian coastline. *American Journal of Applied Sciences* 12 (7):506–515.

Malila, W. A. 1980. Change vector analysis: An approach for detecting forest changes with Landsat. Paper read at *6th International Symposium on Machine Processing of Remotely Sensed Data*, 3–6 June 1980, West Lafayette, IN, USA.

Malpica, J. A. 2007. Hue adjustment to IHS pan-sharpened IKONOS imagery for vegetation enhancement. *IEEE Geoscience and Remote Sensing Letters* 4 (1):27–31.

Man, Q., P. Dong, and H. Guo. 2015. Pixel- and feature-level fusion of hyperspectral and LiDAR data for urban land-use classification. *International Journal of Remote Sensing* 36 (6):1618–1644.

Marcelino, E. V., A. R. Formaggio, and E. E. Maeda. 2009. Landslide inventory using image fusion techniques in Brazil. *International Journal of Applied Earth Observation and Geoinformation* 11 (3):181–191.

McRoberts, R. E. 2014. Post-classification approaches to estimating change in forest area using remotely sensed auxiliary data. *Remote Sensing of Environment* 151:149–156.

Meng, J., X. Du, and B. Wu. 2013. Generation of high spatial and temporal resolution NDVI and its application in crop biomass estimation. *International Journal of Digital Earth* 6 (3):203–218.

Meyer, F. J., D. B. McAlpin, W. Gong et al. 2015. Integrating SAR and derived products into operational volcano monitoring and decision support systems. *ISPRS Journal of Photogrammetry and Remote Sensing* 100:106–117.

Milisavljevic, N. and I. Bloch. 2008. Possibilistic versus belief function fusion for antipersonnel mine detection. *IEEE Transactions on Geoscience and Remote Sensing* 46 (5):1488–1498.

Milisavljević, N. and I. Bloch. 2010. How can data fusion help humanitarian mine action? *International Journal of Image and Data Fusion* 1 (2):177–191.

Muad, A. M. and G. M. Foody. 2012. Super-resolution mapping of lakes from imagery with a coarse spatial and fine temporal resolution. *International Journal of Applied Earth Observation and Geoinformation* 15:79–91.

Mulla, D. J. 2013. Twenty five years of remote sensing in precision agriculture: Key advances and remaining knowledge gaps. *Biosystems Engineering* 114 (4):358–371.

Mwaniki, M. W., M. S. Matthias, and G. Schellmann. 2015. Application of remote sensing technologies to map the structural geology of central region of Kenya. *IEEE Journal of Selected Topics in Applied Earth Observations and Remote Sensing* 8 (4):1855–1867.

Netzband, M. and C. Jürgens. 2010. Urban and suburban areas as a research topic for remote sensing. In *Remote Sensing of Urban and Suburban Areas*, edited by T. Rashed and C. Jürgens. Netherlands: Springer, pp. 1–9.

Nichol, J and M. S. Wong. 2005. Satellite remote sensing for detailed landslide inventories using change detection and image fusion. *International Journal of Remote Sensing* 26 (9):1913–1926.

Otukei, J. R., T. Blaschke, and M. Collins. 2015. Fusion of TerraSAR-x and Landsat ETM+ data for protected area mapping in Uganda. *International Journal of Applied Earth Observation and Geoinformation* 38:99–104.

Paganelli, F., E. C. Grunsky, J. P. Richards, and R. Pryde. 2003. Use of RADARSAT-1 principal component imagery for structural mapping: A case study in the Buffalo Head Hills area, northern central Alberta, Canada. *Canadian Journal of Remote Sensing* 29 (1):111–140.

Paine, J. G., J. Andrews, K. Saylam et al. 2013. Determining wetlands distribution, lake depths and topography using airborne LiDAR and imagery on the North Slope, Deadhorse area, Alaska. In *Final Technical Report: Bureau of Economic Geology*, University of Texas at Austin.

Pal, S. K., T. J. Majumdar, and A. K. Bhattacharya. 2007. ERS-2 SAR and IRS-1C LISS III data fusion: A PCA approach to improve remote sensing based geological interpretation. *ISPRS Journal of Photogrammetry and Remote Sensing* 61 (5):281–297.

Pattetti, S. B, T. J. Majumdar, and A. K. Bhattacharya. 2014. Study of spectral signatures for exploration of Bauxite ore deposits in Panchpatmali, India. *Geocarto International* 30 (5):545–559.

Pitts, R. K., E. Karpouzli, and T. J. Malthus. 2004. Remote sensing techniques for detecting change in coral reef environments: A comparative analysis. In *ASPRS 2004 Annual Conference 'Mountains of Data—Peak Decisions'*. Denver, Colorado: ASPRS.

Playle, N. 2006. Detection of landmines using hyperspectral imaging. In *Detection and Remediation Technologies for Mines and Minelike Targets XI*, edited by J. T. Broach, Harmon, Russell S., Holloway, John H. Jr. Orlando, Florida, USA: SPIE.

Pohl, C. 1996. Geometric aspects of multisensor image fusion for topographic map updating in the humid tropics. Dissertation, Institute of Photogrammetry, Leibnitz University Hannover, Hannover, Germany.

Pohl, C. 2014. Mapping palm oil expansion using SAR to study the impact on the CO_2 cycle. *IOP Conference Series: Earth and Environmental Science* 20 (1):012012.

Pohl, C. 2016. Multisensor image fusion guidelines in remote sensing. In *9th International Symposium on Digital Earth (ISDE 9)*, edited by Hugh Millward, Dirk Werle and Danika van Proosdij. Halifax, Canada: IOP Conference Series: Earth and Environmental Science.

Pohl, C. and M. Hashim. 2014. Increasing the potential of Razaksat images for map-updating in the Tropics. In *8th International Symposium on Digital Earth (ISDE 8)*, Mazlan Hashim, Samsudin Ahmad, and Yin Chai Wang. Kuching, Sarawak, Malaysia: IOP Conference Series: Earth and Environmental Science.

Pohl, C., B. N. Koopmans, and Y. Wang. 1995. The 1995 flood in the Netherlands. In *New Views of the Earth: The Operational Achievements of ERS-1*, edited by ESA. Paris, France: ESA.

Pohl, C. and J. L. van Genderen. 1999. Multi-sensor image maps from SPOT, ERS and JERS. *Geocarto International* 14 (2):34–41.

Pohl, C. and J. L. van Genderen. 2014. Remote sensing image fusion: An update in the context of Digital Earth. *International Journal of Digital Earth* 7 (2):158–172.

Pour, A. B. and M. Hashim. 2014. Integrating PALSAR and ASTER data for mineral deposits exploration in tropical environments: A case study from Central Belt, Peninsular Malaysia. *International Journal of Image and Data Fusion* 6 (2):170–188.

Pour, A. B., M. Hashim, and J. L. van Genderen. 2013. Detection of hydrothermal alteration zones in a tropical region using satellite remote sensing data: Bau goldfield, Sarawak, Malaysia. *Ore Geology Reviews* 54:181–196.

Prakash, A., E. J. Fielding, R. Gens, J. L. van Genderen, and D. L. Evans. 2001. Data fusion for investigating land subsidence and coal fire hazards in a coal mining area. *International Journal of Remote Sensing* 22 (6):921–932.

Prakash, A. and R. Gens. 2011. Chapter 14—Remote sensing of coal fires. In *Coal and Peat Fires: A Global Perspective*, edited by G. B. Stracher, A. Prakash, and E. V. Sokol. Amsterdam: Elsevier.

Rahman, M. M., J. Tetuko Sri Sumantyo, and M. F. Sadek. 2010. Microwave and optical image fusion for surface and sub-surface feature mapping in Eastern Sahara. *International Journal of Remote Sensing* 31 (20):5465–5480.

Raizer, V. 2013. Multisensor data fusion for advanced ocean remote sensing studies. In *IEEE International Geoscience and Remote Sensing Symposium (IGARSS)*, Melbourne, Australia.

Robledo, L., M. Carrasco, and D. Mery. 2009. A survey of land mine detection technology. *International Journal of Remote Sensing* 30 (9):2399–2410.

Roy, P., A. Guha, and K. V. Kumar. 2015. An approach of surface coal fire detection from ASTER and Landsat-8 thermal data: Jharia coal field, India. *International Journal of Applied Earth Observation and Geoinformation* 39:120–127.

Sabins, F. F. 1999. Remote sensing for mineral exploration. *Ore Geology Reviews* 14 (3–4):157–183.

Salentinig, A. and P. Gamba. 2015. Combining SAR-based and multispectral-based extractions to map urban areas at multiple spatial resolutions. *IEEE Geoscience and Remote Sensing Magazine* 3 (3):100–112.

Sango, D., S. Kusano, T. Shibayama, and K. Yoshikawa. 2015. A study for the methodology to monitor Mt. Shinmoe-Dake volcano by combining ALOS-2, Radarsat-2, and TerraSAR-X satellites' data. In *35th Asian Conference on Remote Sensing*, edited by AARS. Manila, Philippines: AARS.

Santurri, L., R. Carlà, F. Fiorucci et al. 2010. Assessment of very high resolution satellite data fusion techniques for landslide recognition. In *ISPRS TC VII Symposium—100 Years ISPRS*, edited by W. Wagner and B. Székely, Vienna, Austria, July 5–7, 2010, IAPRS, Vol. XXXVIII, Part 7B, pp. 492–497.

Scheerer, K. 1997. Airborne multisensor system for the autonomous detection of land-mines. In *Detection and Remediation Technologies for Mines and Minelike Targets II*, edited by A. C. Dubey and R. L. Barnard. Orlando, Florida, USA: SPIE.

Schetselaar, E. M., M. Tiainen, and T. Woldai. 2008. Integrated geological interpretation of remotely sensed data to support geological mapping in Mozambique. *Geological Survey of Finland* 48:35–63.

Schneider, A., M. A. Friedl, and D. Potere. 2009. A new map of global urban extent from MODIS satellite data. *Environmental Research Letters* 4 (4):044003.

Schneider, A., M. Friedl, and C. E. Woodcock. 2003. Mapping urban areas by fusing multiple sources of coarse resolution remotely sensed data. Paper read at *Geoscience and Remote Sensing Symposium, 2003. IGARSS'03. Proceedings. 2003 IEEE International*, Toulouse, France.

Shaban, M. A. and O. Dikshit. 2002. Evaluation of the merging of SPOT multispectral and panchromatic data for classification of an urban environment. *International Journal of Remote Sensing* 23 (2):249–262.

Sheldon, S., X. Xiao, and C. Biradar. 2012. Mapping evergreen forests in the Brazilian Amazon using MODIS and PALSAR 500-m mosaic imagery. *ISPRS Journal of Photogrammetry and Remote Sensing* 74:34–40.

Shen, H., X. Li, Q. Cheng et al. 2015. Missing information reconstruction of remote sensing data: A technical review. *IEEE Geoscience and Remote Sensing Magazine* 3 (3):61–85.

Shupeng, C. and J. L. van Genderen. 2008. Digital Earth in support of global change research. *International Journal of Digital Earth* 1 (1):43–65.

Singh, A. 1989. Review Article. Digital change detection techniques using remotely-sensed data. *International Journal of Remote Sensing* 10 (6):989–1003.

Smith, K. and D. N. Petley. 2009. *Environmental Hazards: Assessing Risk and Reducing Disaster*. New York: Routledge.

Song, Z. and C. Kuenzer. 2014. Coal fires in China over the last decade: A comprehensive review. *International Journal of Coal Geology* 133:72–99.

Stefanski, J., O. Chaskovskyy, and B. Waske. 2014. Mapping and monitoring of land use changes in post-Soviet western Ukraine using remote sensing data. *Applied Geography* 55:155–164.

Stracher, G. B., A. Prakash, and E. V. Sokol. 2013. Coal-geology and combustion. In *Coal and Peat Fires: A Global Perspective*, edited by G. B. Stracher, A. Prakash, and E. V. Sokol. Boston: Elsevier.

Teggi, S. and F. Despini. 2014. Estimation of subpixel MODIS water temperature near coastlines using the SWTI algorithm. *Remote Sensing of Environment* 142:122–130.

Teruiya, R. K., W. R. Paradella, A. R. Dos Santos, R. Dall'Agnol, and P. Veneziani. 2008. Integrating airborne SAR, Landsat TM and airborne geophysics data for improving geological mapping in the Amazon region: The Cigano Granite, Carajás Province, Brazil. *International Journal of Remote Sensing* 29 (13):3957–3974.

Tewes, A., F. Thonfeld, M. Schmidt et al. 2015. Using RapidEye and MODIS data fusion to monitor vegetation dynamics in semi-arid rangelands in South Africa. *Remote Sensing* 7 (6):6510–6534.

Tilley, D. G. 1992. SAR-optical image fusion with Landsat, Seasat, Almaz and ERS-1 satellite data. In *XVIIth ISPRS Congress*. Washington DC: ISPRS.

Tilley, D. G., Y. V. Sarma, and R. C. Beal. 1992. Ocean data reduction and multi-sensor fusion studies of the Chesapeake region. Paper read at *International Geoscience and Remote Sensing Symposium (IGARSS'92)*, 26–29 May 1992, Houston, Texas.

Tobin, G. A. and B. E. Montz. 1997. *Natural Hazards: Explanation and Integration*. New York, USA: Guilford Press.

Tralli, D. M., R. G. Blom, V. Zlotnicki, A. Donnellan, and D. L. Evans. 2005. Satellite remote sensing of earthquake, volcano, flood, landslide and coastal inundation hazards. *ISPRS Journal of Photogrammetry and Remote Sensing* 59 (4):185–198.

Tuell, G. 1998. The use of high resolution airborne synthetic aperture radar (SAR) for shoreline mapping. In *International Archives of Photogrammetry and Remote Sensing*. Columbus, Ohio: ISPRS.

Umbert, M., S. Guimbard, G. Lagerloef et al. 2015. Detecting the surface salinity signature of Gulf Stream cold-core rings in Aquarius synergistic products. *Journal of Geophysical Research-Oceans* 120 (2):859–874.

Umbert, M., N. Hoareau, A. Turiel, and J. Ballabrera-Poy. 2014. New blending algorithm to synergize ocean variables: The case of SMOS sea surface salinity maps. *Remote Sensing of Environment* 146:172–187.

UN. 2014. World urbanization prospects, the 2014 revision. United Nations, Department of Economic and Social Affairs.

UN-ISDR. 2005. Hyogo framework for action 2005–2015: Building the resilience of nations and communities to disasters. Paper read at *Final report of the World Conference on Disaster Reduction (A/CONF. 206/6)*, Kobe, Hyogo, Japan.

van der Meer, F. D., H. M. A. van der Werff, F. J. A. van Ruitenbeek et al. 2012. Multi- and hyperspectral geologic remote sensing: A review. *International Journal of Applied Earth Observation and Geoinformation* 14 (1):112–128.

Verhoeven, G. 2015. TAIFU: Toolbox for archaeological image fusion. Paper read at *AARG 2015*, Santiago de Compostela, Spain.

Viera, C. A. O., P. M. Mather, and P. Aplin. 2002. Multitemporal classification of agricultural crops using the spectral-temporal response surface. In *Analysis of Multi-Temporal Remote Sensing Images*, edited by Bruzzone, L. and Smits, P. London: World Scientific.

Villa, G., J. Moreno, A. Calera et al. 2013. Spectro-temporal reflectance surfaces: A new conceptual framework for the integration of remote-sensing data from multiple different sensors. *International Journal of Remote Sensing* 34 (9–10):3699–3715.

Walker, J. J., K. M. de Beurs, and R. H. Wynne. 2014. Dryland vegetation phenology across an elevation gradient in Arizona, USA, investigated with fused MODIS and Landsat data. *Remote Sensing of Environment* 144:85–97.

Wang, L., Y. Tian, X. Yao, Y. Zhu, and W. Cao. 2014. Predicting grain yield and protein content in wheat by fusing multi-sensor and multi-temporal remote-sensing images. *Field Crops Research* 164 (1):178–188.

Werner, A., C. D. Storie, and J. Storie. 2014. Evaluating SAR-optical image fusions for urban LULC classification in Vancouver Canada. *Canadian Journal of Remote Sensing* 40 (4):278–290.

Wielaard, N. 2011. *Impact of Oil Palm Plantations on Peatland Conversion in Sarawak 2005–2010*. Report by Sarvision for Wetlands International, 16 pp. http://www.forestcarbonasia.org/other-publications/impact-of-oil-palm-plantations-on-peatland-conversion-in-sarawak-2005-2010/.

WorldBank. *Agriculture, Value Added (% of GDP)*. World Bank 2015 [cited 29 October 2015]. Available from http://data.worldbank.org/indicator/NV.AGR.TOTL.ZS/countries/1W?display = map.

Wu, M. Q., C. Y. Wu, W. J. Huang, Z. Niu, and C. Y. Wang. 2015. High-resolution leaf area index estimation from synthetic Landsat data generated by a spatial and temporal data fusion model. *Computers and Electronics in Agriculture* 115:1–11.

Yan, W. Y., A. Shaker, and N. El-Ashmawy. 2015. Urban land cover classification using airborne LiDAR data: A review. *Remote Sensing of Environment* 158:295–310.

Yang, C., J. Everitt, and J. Bradford. 2009. Evaluating high resolution SPOT 5 satellite imagery to estimate crop yield. *Precision Agriculture* 10 (4):292–303.

Yang-Lang, C., L. S. Liang, H. Chin-Chuan, F. Jyh-Perng, L. Wen-Yew, and C. Kun-Shan. 2007. Multisource data fusion for landslide classification using generalized positive Boolean functions. *IEEE Transactions on Geoscience and Remote Sensing* 45 (6):1697–1708.

Yésou, H., Y. Besnus, J. Rolet, J. C. Pion, and A. Aing. 1993. Merging Seasat and SPOT imagery for the study of geological structures in a temperate agricultural region. *Remote Sensing of Environment* 43 (3):265–279.

Yu, L., Y. Nie, F. Liu, J. Zhu, Y. Yao, and H. Gao. 2012. Research on data fusion method for archaeological site identification. Paper read at *Proceedings of 2012 International Conference on Image Analysis and Signal Processing, IASP 2012*, Hangzhou, China.

Zeng, Y., J. Zhang, J. L. van Genderen, and Y. Zhang. 2010. Image fusion for land cover change detection. *International Journal of Image and Data Fusion* 1 (2):193–215.

Zhang, C. 2015. Applying data fusion techniques for benthic habitat mapping and monitoring in a coral reef ecosystem. *ISPRS Journal of Photogrammetry and Remote Sensing* 104:213–223.

Zhang, L., X. Ding, and Z. Lu. 2015. Ground deformation mapping by fusion of multi-temporal interferometric synthetic aperture radar images: A review. *International Journal of Image and Data Fusion* 6 (4):289–313.

Zhang, X. M., C. J. S. Cassells, and J. L. van Genderen. 1998. Multi-sensor data fusion for the detection of underground coal fires. *Geologie en Mijnbouw* 77 (2):117–128.

Zhang, Y. 2001. Detection of urban housing development by fusing multisensor satellite data and performing spatial feature post-classification. *International Journal of Remote Sensing* 22 (17):3339–3355.

Zhou, Y. and F. Qiu. 2015. Fusion of high spatial resolution WorldView-2 imagery and LiDAR pseudo-waveform for object-based image analysis. *ISPRS Journal of Photogrammetry and Remote Sensing* 101:221–232.

Zhukov, B., D. Oertel, F. Lanzl, and G. Reinhackel. 1999. Unmixing-based multisensor multiresolution image fusion. *IEEE Transactions on Geoscience and Remote Sensing* 37 (3):1212–1226.

Zia, A., V. DeBrunner, A. Chinnaswamy, and L. DeBrunner. 2002. Multi-resolution and multi-sensor data fusion for remote sensing in detecting air pollution. Paper read at *Proceedings of Fifth IEEE Southwest Symposium on Image Analysis and Interpretation, 2002*, Santa Fe, NM, USA.

7

Conclusions and Trends

In this last chapter of the book, we describe some of the most important issues in the field of remote sensing image fusion. We explain the remaining challenges and several current and future trends in this field and highlight some of the remaining research questions that need to be answered over the coming years.

7.1 Challenges

Taking into account the increased demand for details and parameters to be considered, researchers are facing new challenges, such as the following:

- Integration of more than three spectral bands.
- High-resolution images require higher measures in geometric accuracy prior to image fusion.
- Observation angles impose different shadow features in different images in high-resolution data creating artifacts in fused images.
- Changes in the range of spectral bands of new-generation sensors require particular consideration in matching the panchromatic channel.
- The high spatial resolution imposes new features being disturbed (e.g., blurring of building edges or moving cars) that were not even noticeable before.
- Higher spectral and spatial resolutions put higher demands on the quality of fused images, and therefore on the fusion technique itself.
- The geometric and radiometric requirements for fused imagery that is of good quality put a high demand on the operator's knowledge and processing complexity.

7.2 Trends

Back in the late 1990s/early 2000s, one of the main research areas in remote sensing image fusion was in the field of pansharpening, by combining high- and low-resolution images from the same satellite system. Examples are the combination of 10 m panchromatic images with the 20 m multispectral images from the SPOT satellite (Vrabel 1996). Other commonly used image combinations used optical data from different platforms, such as SPOT 10 m panchromatic with Landsat multispectral images at 30 m spatial resolution (Chavez et al. 1991; Price 1999; Ranchin and Wald 2000). Other processing approaches focused on the use of optical and microwave remote sensing images due to applications, such as geology that required input from SAR (Harris and Murray 1990; Yésou et al. 1993). In the meantime, image fusion has become well established. Pansharpening still enjoys high popularity. There are several comprehensive review papers, including an overview on the different applications and recommendations (Zhang 2008; Ehlers et al. 2010; Aiazzi et al. 2012; Zhang and Mishra 2014). The importance of pansharpening is visible through a patent on the Gram–Schmidt (GS) pansharpening algorithm that has been filed in the United States, No. 6011875 (Laben et al. 2000) and commercially used algorithms, (Ehlers 2004; Zhang and Mishra 2014). Now, many fusion algorithms have become standard functions in most commercial software packages such as ERDAS, PCI, ENVI, etc. Satellite- and airborne sensors with higher spatial and spectral resolution are increasingly available. As a result, remote sensing image fusion is now also feasible between satellite imagery and airborne data sets such as hyperspectral (Garzelli et al. 2010; Palsson et al. 2014; Dalla Mura et al. 2015; Loncan et al. 2015) and LiDAR data (Berger et al. 2013a,b; Dalla Mura et al. 2015; Wu and Tang 2015).

Trends move toward hybrid approaches; hybrid in terms of fusion levels, fusion techniques, and object-based image analysis (OBIA) in general. The access to high-resolution imagery led to the need for OBIA techniques that have rapidly evolved over the last decade. The classification is applied on features rather than on pixels. The feature extraction is implemented by segmentation to identify spatially adjacent pixels with homogeneous criteria (Zhou and Qiu 2015). Fusion in the context of classification relies on more and more sophisticated and intelligent processing. After the introduction of decision trees, these are further developed into decision forests, which consist of several decision trees. This significantly increases the accuracy of classification results. An example is the random forest (RF) algorithm, which outperforms other learning methods (Rokach 2016). Hence, as can be seen from the above, the overall trend in remote sensing image fusion has been from relatively simple pansharpening through various stages of complexity to feature-based fusion, then decision-based fusion, to data fusion, 3D fusion, information fusion, to advanced sensor fusion. This is illustrated in Figure 7.1.

The development can be sketched as follows:

1. Optical pixel-based fusion, for example, pansharpening
2. Pixel-based fusion of optical data with data from other wavelengths, for example, VIR/SAR, VIR/TIR
3. Feature-based fusion, for example, detection of specific objects, such as landmines and landslides
4. Decision-based fusion, for example, different classification results
5. Data fusion, for example, fusing remote sensing images with data from a GIS and other spatial data sets
6. 3D fusion, for example, fusing stereo satellite data with DTMs, or LiDAR point clouds with stereo digital aerial imagery
7. Information fusion, for example, fusing remote sensing images with statistics, sound, or other ancillary sources of information
8. Advanced sensor fusion, for example, the Internet of Things (IoT), fusing *in situ* sensor webs with airborne and spaceborne sensors,

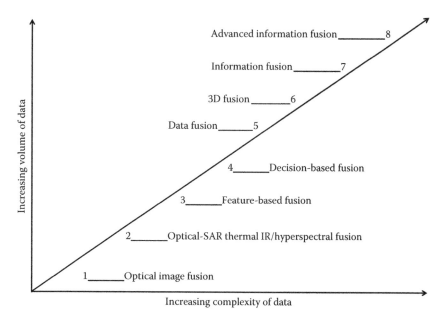

FIGURE 7.1
A fusion roadmap: The Internet of Things.

with other GPS location-based devices (ubiquitous positioning) to locate people and objects, plus abilities to monitor (and control) distant objects, that is, physical world connected to web world

The rest of this chapter will describe some of the new technologies that will influence remote sensing fusion over the coming years. These aspects include data mining, cloud computing, Big Data, and IoT.

7.3 Data Mining

With the ever-increasing size of remote sensing fused data sets, it becomes increasingly important to extract meaningful information from such fused data sets. One method becoming increasingly useful to tackle this issue is "data mining." This is the computational process of discovering patterns in large remote sensing data sets. The main goal of the data mining process is to extract information from the fused data set and transform it into an understandable structure for further interpretation and analysis. The idea is to extract previously unknown, interesting patterns for anomaly detection in the fused remote sensing data (Mitsa 2010; Kantardzic 2011). With the growing interest in remote sensing image and data fusion, and the increasing complexity of high-resolution and hyperspectral imaging systems, data mining on high-dimensional heterogeneous imagery has become a vital aspect in image fusion applications. The roles of remote sensing image fusion and data mining are very complementary, in that the data-mining-driven process of automatic object discovery can be integrated into the image fusion driven target or object identification activity. Data mining techniques are increasingly used to examine remote sensing fused data for applications in areas such as land cover classification (Pugh et al. 2006), forestry (Leckie 1990), mapping (Parsons and Carpenter 2003), and, especially, for examining large temporal data sets (Mitsa 2010).

The main approach is first to produce a very large layered stack of co-registered, processed, multispectral/multi-sensor imagery plus any other available layers of contextual information such as 3D terrain models to produce features. Then by data mining techniques, such as neural networks, one can make a vertical cut through such a stack, which will then correspond to a feature vector at each pixel, as illustrated in Figure 7.2. In this way, the neural network can use this training data to search for more of the desired target or class of land cover.

Using a very large remote sensing data set, consisting of a multi-season Landsat and Radarsat imagery taken during three different seasons, together with a 10 m digital elevation model, Pugh et al. (2006) fused all this data using neural image fusion approaches. After this, they carried out the data

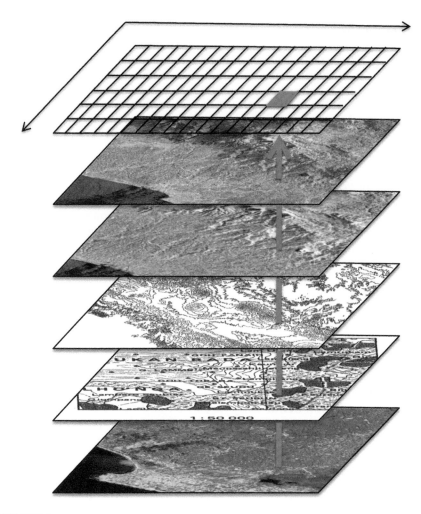

FIGURE 7.2
Data mining through a large fused remote sensing data set.

mining to search for new patterns and enhanced land cover classification. Additional image layers can help to exploit any spatial and contextual patterns data in the vicinity of any pixel. By adding the digital elevation data, three-dimensional layers provided extra information on terrain height, slope, and mean curvature. Once such a large fused data set was produced, this can be interactively mined to detect objects at the pixel level from their combined multispectral and multi-sensor signatures in conjunction with their local characteristics. Data mining techniques that have been used in the context of remote sensing image fusion include maximum likelihood classifiers, neural networks, decision trees, and support vector machines (SVM). Milenova and Campos (2005) used such an approach to carry out data mining on a fused hyperspectral data set from the airborne visible infrared imaging spectrometer (AVIRIS) sensor. They used the 185 channels of data for their data fusion and mining work. Such a data mining approach is most suitable for a very high-dimensional feature space, such as large multi-temporal data sets for change detection and hyperspectral imagery for land use applications. One of the best algorithms for data mining of such high-dimensional fused data sets is SVM, which outperforms more traditional approaches such as maximum likelihood classifiers (Chiarella et al. 2003). It is suitable irrespective of the dimensionality of the fused data set (Shah et al. 2003).

Without having to replicate all one's data, the database views can capture different slices of the fused data. For example, fused temporal imagery, images of any of the input sensors used for the image fusion, can be used directly for model generation and subsequent data analysis. Thus, data mining methods allow one to inspect the fused data sets, either directly, or via visualization tools. This will be discussed in more detail in Sections 7.4 and 7.5.

7.4 Cloud Computing

As we have shown in Chapter 6, remote sensing image fusion has a very wide scope of applications. RSIF products have become an indispensable source of information in our daily life, whether it be for urban planning, agriculture, environmental management, mineral exploration, or climate change (Robila 2006). Until recently, it was still possible to fuse remote sensing data because satellite and other platforms were available but with limited spatial, spectral, and temporal resolutions. However, this picture has changed dramatically because of the large number of satellites now available to the Earth observation community, at ever-increasing spatial resolution, with more and more spectral bands and with a very high revisit time. Although these advances aid in our understanding of Earth as an integrated system, the storage capacity and computer processing power required to analyze such large data sets are beyond the means of individual desktop computers or workstations. In order to fuse and analyze such large remote sensing data sets, cloud computing has provided a solution.

Cloud computing provides a simple way to access servers, storage, databases, and a broad set of application services over the Internet. It is a powerful tool to perform large-scale and complex computing (Hashem et al. 2015). The term has emerged from the use of a "cloud" as a symbol for a communications network (compare with Figure 7.3). This can be any kind of network, that is, local, national, or global. The cloud often refers to the

FIGURE 7.3
Cloud computing of big data in a geospatial context.

Internet where a data center provides the servers. The cloud comprises not only the connection between systems but also servers that perform the required data processing. Cloud computing providers such as Amazon or Google own and maintain the network-connected hardware required for these application services, while users have the provision and use what they need via a web application. A definition of cloud computing is provided by the National Institute of Standards and Technology (NIST) in the United States. It states that cloud computing "is a model for allowing ubiquitous, convenient, and on-demand network access to a number of configured computing resources (e.g., networks, server, storage, application, and services) that can be rapidly provisioned and released with minimal management effort or service provider interaction" (Jansen and Grance 2011).

Hence, cloud computing is considered to be a convenient solution to the problem of fusing very large Earth observation data sets, since the "cloud" provides unlimited storage capacity and elastic computer services in the form of the following three platform models:

- Software as a service (SAAS)
- Platform as a service (PAAS)
- Infrastructure as a service (IAAS)

Additionally, cloud computing can reduce costs significantly, since it is now possible to pay and use on demand, instead of having to purchase advanced hardware and software to process these large remote sensing data sets (Ghaffar and Tuong Thuy 2015).

Many of the large space agencies such as NASA and ESA are moving their remote sensing data to the cloud. In Europe, for example, the HELIX Nebula Partnership is a new, pioneering partnership between leading IT providers and some of Europe's largest research centers to chart a course for sustainable cloud services for research communities. ESA is creating an Earth observation platform called Helix Nebula, focusing on earthquake and volcano research on the cloud. In partnership with the Centre National d'Études Spatiales (CNES) in France and the National Aeronautics and Space Research Centre (DLR), they are collaborating with the National Research Council (CNR) in Italy, to create an Earth observation platform focusing on earthquake and volcano research (Helix Nebula 2014). Such projects bring the global geo-hazard community onto a common platform that enables the correlation, processing, and fusion of observation data for supersites monitoring.

In the United States, the NASA NEX is a collaboration and analytical platform that combines state-of-the-art supercomputing, Earth system modeling, workflow management, and NASA remote sensing data. Through NEX, users can explore and analyze large Earth science data sets, run and share modeling algorithms, collaborate on new or existing projects, and exchange workflows and results within and among other science communities. Three NASA NEX data sets are now available to all via Amazon S3, a cloud service by Amazon Web Services. One data set, the NEX downscaled climate simulations, provides high-resolution climate change projections for the 48 contiguous U.S. states. The second data set, provided by the MODIS instrument on NASA's Terra and Aqua satellites, offers a global view of Earth's surface every 1 to 2 days. Finally, the Landsat data record from the U.S. Geological Survey provides the longest existing continuous space-based record of Earth's land. These remote sensing data sets allow the user to fuse the imagery for making detailed change detection studies of land cover, and many other topics, and are especially useful for long-term study of climate change (e.g., desertification, deforestation, and land degradation) (Amazon 2015).

A good example of cloud computing in relation to RSIF is Google's "Earth Engine." It is a cloud-computing platform for processing large sets of satellite imagery and other large remote sensing data sets. Using fusion techniques, it has created a global temporal fusion, or time-lapse image of the world, which allows the user to view landscape changes in any location on Earth (Kluger 2015). This involved using 30 years of Landsat data and some two million images. To produce it, it needed 2 million hours using 66,000 computers. It took one and a half days to prepare.

7.5 Big Data

The term "Big Data" first appeared in 2001 from Doug Laney of META Group (Baumann et al. 2015). Big Data is high-volume, high-velocity, and/or high-variety information assets that demand cost-effective, innovative forms of information processing that enable enhanced insight, decision making, and process automation (Gartner 2013). It is anticipated that the digital universe will grow by a factor of 10 in the period 2013–2020, from 4.4Zb to 44Zb. This means that it doubles every 2 years (Turner et al. 2014). Google currently processes 25Pb/day (Lee and Kang 2015). The data to be handled in the context of climate change is going to reach 350Pb by 2030 (Huadong 2015). All this has been enabled by the development of communication and digital storage technologies. There is a need for a proper approach to access, process, and maintain this data. The challenges in Big Data are volume, velocity, and variety as the first "3Vs," complemented by veracity leading to "4Vs" (Baumann et al. 2015). Accuracy, reliability, and adequate metadata are current research fields in the context of Big Data. Big Data handling allows the extraction of information from large remote sensing and complex image data sources for many applications that could benefit society and the sustainability of our planet. This requires the integration/fusion of increasingly diverse image data sets and ancillary data sources, the design of new forms of data collection, extraction, and interpretation of knowledge utilizing sophisticated data fusion algorithms, complex simulation models, and several other currently computer intensive approaches. Various modalities are processed simultaneously, often covering very large areas. This leads to the need for distributed computing, that is, cloud computing to reduce processing time (Gomez-Chova et al. 2015). By its nature, geospatial data tends to be big, especially when looking at it from a global perspective. Cloud computing is one of the main reasons why it has become possible to fuse very large data sets of Earth observation data, such as temporal fusion of 40 years of Landsat images, of large areas of Earth for climate change studies. It is often noted that for RSIF projects, 80% of the analyst's time is spent on data preparation, prior to the fusion process, and only 20% is spent on looking for insights to analyze the fused products. With Big Data technology, data preparation tasks such as geometric corrections, geocoding, and co-registration of various imagery types are becoming easier, with less technology and infrastructure required to support the fusion process (see Figure 7.4).

Variety makes Big Data really big. Big Data comes from a great variety of remote sensing data sources. In terms of remote sensing images, Table 7.1 provides a glimpse on data volume to be handled, just looking at the original data. The processing and exploitation will then add volume to derive the necessary conclusions. Temporal changes, over a 40-year period enable detailed studies of global change, deforestation, urban growth, land

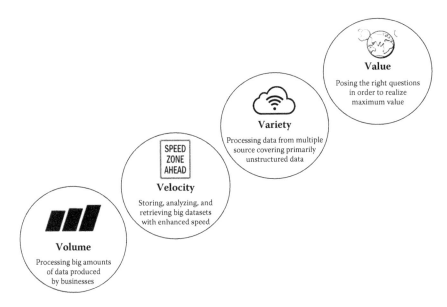

FIGURE 7.4
4Vs of Big Data.

degradation, natural disasters such as floods, etc. to be analyzed in a timely manner. This can be done at global scales, and many animations exist showing changes in land cover, and other environmental parameters of the entire world on a monthly, weekly, and even daily basis. When looking at such animations, one does not realize the huge mosaicking, geometric, and radiometric corrections that were necessary to make them possible. The same is true for the billions of users of Google Earth, where hundreds of thousands of high-resolution images have been radiometrically and geometrically processed into one seamless fused image of the world. The volume or size of data now is larger than terabytes and petabytes. The large volume and types of Earth observation data, with an ever-increasing number of optical, thermal-infrared, SAR, and hyperspectral satellite sensors, with higher temporal

TABLE 7.1

Data Volume of Common Remote Sensing Images

Type	File Size
WorldView-2 (16 bit)	2.5 Gb
QuickBird (16 bit)	1.5 Gb
Landsat 8	2 Gb
Landsat 7	500 Mb
Landsat 4-5 TM	350 Mb
Landsat 1-3 MSS	50 Mb

frequency, outstrips traditional storage, and image processing and analysis techniques.

Previously, it was not so easy to fuse such large data sets, as, besides the size of the actual satellite images, with many bands, there are all the intermediate products that need large storage space, such as the geometric correction, radiometric corrections for image registration, etc. But, now, with cloud technology, and Big Data analytics, we can readily fuse such large image data sets.

Figure 7.5 is an example of a fused image of a satellite. The imagery used to produce this fused product was acquired on February 16, 2012. It is an image acquired by the Chinese ZY-3 stereo mapping satellite, which has acquired more than one petabyte of data since it was launched in January 2011. It provides a panchromatic image at 2.1 m spatial resolution as well as four multispectral bands at 5.8 m. The imagery was provided courtesy of the Chinese Academy of Surveying and Mapping in Beijing, China. The images were all fused using the PANSHARP (UNB based) image fusion algorithm provided by the PCI Geomatica software.

Velocity is required not only for Big Data but also for many remote sensing applications such as disaster monitoring. There are several approaches to collecting, storing, processing, and analyzing Big Data. Big Data comes from a great variety of sources and in several types such as structured, semistructured, and unstructured. Structured data inserts a data warehouse already tagged and easily sorted, but unstructured data is random and more difficult to analyze.

The volume or the size of data now is larger. Unstructured data refers to information that either does not have a predefined data model or does not fit well into relational tables. Unstructured data are the fastest-growing type of data; some examples are Earth observation satellite imagery, remote sensors from airborne, UAV or other platforms, video, GIS files, and statistical data files. There are several techniques to address this problem space of unstructured analytics. The techniques share common characteristics of scale-out, elasticity, and high availability. MapReduce, in conjunction with the Hadoop Distributed File System (HDFS) and HBase database, as part of the Apache Hadoop project, are modern approaches to analyze unstructured data. Hadoop clusters are effective means of processing and fusing such massive volumes of data (Bakshi 2012).

The coupling with many new enabling technologies, such as data compression, data mining, cloud computing, artificial intelligence, neural networks, and deep learning, have enabled ever-larger data sets to be prepared and processed. The trend to larger data sets is therefore due to the additional information derivable from analysis of the large sets of available Earth observation data, combined with the multitude of related data sets, such as on environmental parameters, statistical data sets from ground-based sensors, and sensor webs. Thus, using image and data fusion technologies, one can now find correlations, spot trends, determine quality of research, and

FIGURE 7.5
Dubai Tower and Palm Island as seen by ZY-3. (Courtesy of the Chinese Academy of Surveying and Mapping in Beijing, China.)

discover things not visible on any one data set by itself, thus aiding in the study of many temporal changes taking place on Earth. Specifically, the application of Big Data technology in remote sensing has mainly focused on parallel systems (Ma et al. 2015). The major components of such parallel systems include system architecture, parallel file system and parallel I/O, programming models, data managing at a multilevel memory hierarchy, and task scheduling. Hampton et al. (2013) show how in the field of ecology and environment Big Data is contributing to global-scale environmental issues, from climate change and food security to the spread of disease and the availability of clean water. Society is asking ecologists for information that is specific to particular problems, places, and times, and also predictive, prescriptive, and scalable. Thus, Big Data analytics is an area where remote sensing image and data fusion can play an important role over the coming years.

7.6 Internet of Things

The IoT can be defined as a network of physical objects or "things" embedded with many types of sensors, software, and the Internet, which enables such objects to collect and exchange data. According to Evans (2011), the IoT is simply the point in time when more "things or objects" were connected to the Internet than people. That point happened in 2009. Hence, objects on Earth can be sensed by remote sensing satellites, aircraft, drones, *in situ* sensor networks, and then integrated and controlled remotely across an existing network infrastructure. This may create opportunities for more direct integration between the physical world as measured by remote sensing sensors and computer-based systems, resulting in improved efficiency, speed, and accuracy.

Experts estimate that the IoT will consist of about 50 billion devices or objects, wirelessly connected to the IoT by 2020 (Evans 2011). However, this will require a major increase in IP addresses, as the current version of Internet (IPv4) accommodates only 4.3 billion unique addresses. IPv6 will be able to accommodate the extremely large address space required to allow the large number of devices to be connected to the Internet.

"Things" in the IoT can refer to a very wide range of devices, from transponders on farm animals, to monitor devices for wild animal migration patterns, to *in situ* stream gauges to monitor river/lake levels for flood prediction, or to field operation devices that may assist fire fighters in forest fire fighting and search and rescue operations, coupling such devices with airborne or satellite thermal-infrared data, GIS databases, GPS, and forest fire spreading models. Hence, all such devices collect useful data with the help of existing remote sensing technologies, and then the data can autonomously flow the data to and between other devices.

In Figure 7.6, we have placed some applications at the top, and types of users at the bottom. In the middle is the IoT, with all its sensors, *in situ*, embedded, from remote sensing and other sources of connected data, together with Big Data analytics and visualization tools, to provide data of "anything, anywhere, anytime." At the top are some of the main applications at present of IoT, such as in the home (health issues, utilities, security, and appliances), in transport (traffic monitoring, parking, emergency services, highway control, and logisitics), in the community (environmental monitoring such as parks, waste disposal, shopping, and security surveillance), national issues (agriculture, forestry, water resources, disaster monitoring, remote monitoring, and national security), and international aspects, such as global change issues, major international technological cross-border and natural disasters, and sea level rise. Among the users, listed under the figure, we find doctors and other social workers, home users for personal, individual use, local, regional, national, and international policy makers, plus businesses, and military users.

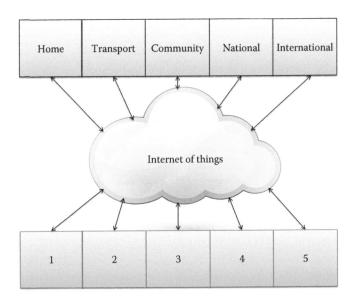

FIGURE 7.6
The Internet of Things, showing end users and application areas (gray = examples of applications; red = users).

The interconnection of these embedded devices can usher in more automation in many fields, while enabling advanced applications like "smart cities," where high-resolution satellite data, airborne LiDAR data, can be coupled to sensors in the city to monitor traffic, garbage collection, environmental pollution, and many other urban problems. One of the requirements to develop the IoT further is that all objects, people, vehicles, etc. will need to be equipped with a unique identifier, so that objects can then be managed by computers. Miniscule identifying devices or machine-readable devices, such as barcodes, QR, can do such tagging codes, near-field communication, digital watermarking, etc.

The IoT finds applications in many fields of Earth observation, such as environmental monitoring, urban planning, transportation, mineral exploration, and natural disaster prediction and monitoring.

In the field of environmental monitoring, the IoT typically will use *in situ* sensor networks plus remote sensing imagery and related models to monitor air quality, water quality, and land surface parameters (Li et al. 2012). Such intelligent environmental sensors, measuring temperature, humidity, air pressure, etc., will serve as the front-end devices to be connected to other networks, to gather and process contextual information from the environment. IoT is also being increasingly used for water network monitoring. Sensors, both *in situ* and from air/spaceborne platforms, are measuring critical water parameters to ensure reliable, high-quality water supply as well as contamination among storm water drains, drinking water, and sewage disposal. Image and data fusion can assist in also extending such networks to monitoring irrigation in agricultural land. Soil parameters can be monitored in a similar fashion and, linked with remote sensing imagery, can allow informed decision making in precision agriculture (Gubbi et al. 2013).

The concept of a "smart city" is another field where many achievements have already been made in using the IoT. Songdo, in South Korea, is already wired up as a smart city. This enables many urban planning functions to be carried out, such as monitoring and analyzing traffic patterns, public transport, parks and open spaces, sports facilities, and any other feature/element needed by the planners with but little human intervention. The image in Figure 7.7 can be fused with any or all of the other urban data sets available via the IoT. For example, critical urban infrastructure such as bridges, railroads, and sensitive industrial installations can be monitored. Any changes in conditions that could compromise safety can be mitigated. It can also be used for scheduling repair and maintenance facilities in the city, by coordinating the tasks between different service providers and users of these facilities. Many other cities around the world are working on similar smart city concepts, based on everything being geocoded and connected in order to better manage large, complex cities.

As we have seen in the previous chapter on applications of remote sensing image fusion, natural and man-made disasters are a major problem for mankind. The IoT can be expected to play an ever-increasing role in alerting us before disasters happen as well as in disaster recovery after the event.

FIGURE 7.7
GoogleEarth Imagery of the Smart City Songdo in South Korea. (Image courtesy 2015 CNES/ Astrium.)

TABLE 7.2

Examples of IoT in Some Earth Observation Domains

Parameters	Smart City	Agriculture/ Forestry	Water Resources	Transport	Natural/ Anthropogenic Disasters	Environment
Network size	Medium/large	Medium/large	Large	Medium/large	Large	Large
Users	Urban planners	Land owners	Government agencies	Local, regional, and national agencies	Land owners	Local, regional, national, and international agencies
	Policy makers General public	Farmers Policy makers	Water boards	General public	Farmers Policy makers	General public
Internet connectivity	Wifi, 3G, 4G	Wifi, satellite communication (GPS, RS)	Satellite communication (GPS, RS), microwave links	Wifi, satellite communications (GPS)	Wifi, 3G, 4G, satellite communications (GPS, RS)	Wifi, 3G, 4G, satellite communications (GPS, RS)
Data management	Shared server	Local server, shared server	Shared server	Shared server	Local server, shared server	Local server, shared server
Bandwidth	Large	Medium	Medium	Medium	Large	Large

Embedded *in situ* sensors, coupled with smart phones for early warning, together with real-time remote sensing data from all sorts of platforms and sensors, will aid in the overall disaster management cycle (Vongsingthong and Smanchat 2014). Examples of IoT in Earth observation are listed in Table 7.2.

In the IoT, the precise geographic location of a thing, as well as the exact geographic dimensions or size of a thing, is very important. On the Internet, information is processed and managed by people, with less emphasis on location in time and space. On the IoT, much of the information will come from embedded sensors in the environment, so that location and scale become very important. Thus, networks of sensors and remotely controlled objects will lead to massively parallel sensor fusion of objects at different scales, using many of the methodologies and algorithms from remote sensing image and data fusion, adapted to the new possibilities offered by the IoT. Most of this is being made possible by the deep penetration of smartphones and advances in wireless communication technology (Vermesan and Friess 2013).

As new display technologies are developed, more and more creative visualizations will allow the users to display the large amount of information (data and imagery) in new, innovative ways. The recent advances in touch screen technologies and the use of smart tablets and phones have especially made such visualizations very intuitive. However, at present, it is rather difficult to produce useful 3D displays for such heterogeneous spatio-temporal data (Post et al. 2012).

7.7 Summary

This final chapter of the book listed some of the main current research issues and several key future research trends in the field of remote sensing image fusion. In terms of the current research issues, these are concerned with dealing with the ever higher spatial and spectral resolution of both satellite and airborne platforms, requiring new approaches to deal with the aspects of off-nadir viewing, fusing more than three bands, the geometric aspects prior to fusion, and several other important parameters. With regard to the new research trends, the field has moved on from simple pansharpening to the fusion of airborne imagery (e.g., LiDAR and hyperspectral data sets) with high or very high spatial resolution Earth observation data. Another major trend is the use of hybrid fusion approaches, in terms of fusion levels and object-based image analysis. Thus, the overall trend in RSIF has been from the relatively basic image fusion through various stages of complexity to feature-based fusion, then decision-based fusion, to data fusion, information fusion, to advanced sensor fusion. The chapter concluded with some of the key technologies, which are already starting to offer many new benefits and opportunities to advance the field of remote sensing image fusion. These include data mining, cloud computing, Big Data analytics, artificial intelligence, neural networks, and the IoT. The inputs from all these disciplines are crucial for advancing the exciting field of remote sensing image and data fusion, which, in our view, will expand rapidly over the coming years, and hence become an ever-increasing vital component of remote sensing.

References

Aiazzi, B., L. Alparone, S. Baronti, A. Garzelli, and M. Selva. 2012. Twenty-five years of pansharpening: A critical review and new developments. In *Signal and Image Processing for Remote Sensing*, edited by C. H. Chen. Boca Raton, FL: CRC Press.

Amazon. *NASA NEX* 2015 [cited 12 November 2015]. Available from http://aws.amazon.com/nasa/nex/

Bakshi, K. 2012. Considerations for big data: Architecture and approach. Paper read at *2012 IEEE Aerospace Conference*, Big Sky, MT, USA, March 3–10, 2012.

Baumann, P., P. Mazzetti, J. Ungar et al. 2015. Big data analytics for earth sciences: The earthserver approach. *International Journal of Digital Earth* 9 (1):3–29.

Berger, C., M. Voltersen, R. Eckardt et al. 2013a. Multi-modal and multi-temporal data fusion: Outcome of the 2012 GRSS data fusion contest. *IEEE Journal of Selected Topics in Applied Earth Observations and Remote Sensing* 6 (3):1324–1340.

Berger, C., M. Voltersen, S. Hese, I. Walde, and C. Schmullius. 2013b. Robust extraction of urban land cover information from HSR multi-spectral and LiDAR data. *IEEE Journal of Selected Topics in Applied Earth Observations and Remote Sensing* 6 (5):2196–2211.

Chavez, P. S., S. C. Sides, and J. A. Anderson. 1991. Comparison of three different methods to merge multiresolution and multispectral data: Landsat TM and SPOT panchromatic. *Photogrammetric Engineering & Remote Sensing* 57 (3):295–303.

Chiarella, M., D. Fay, A. M. Waxman, R. T. Ivey, and N. Bomberger. 2003. Multisensor image fusion and mining: From neural systems to COTS software with application to remote sensing AFE. Paper read at *2003 IEEE Workshop on Advances in Techniques for Analysis of Remotely Sensed Data*, Greenbelt, MD, October 27–28, 2003.

Dalla Mura, M., S. Prasad, F. Pacifici, P. Gamba, J. Chanussot, and J. A. Benediktsson. 2015. Challenges and opportunities of multimodality and data fusion in remote sensing. *Multimodal Data Fusion* 103 (9):1585–1601.

Ehlers, M. 2004. Spectral characteristics preserving image fusion based on Fourier domain filtering. Paper read at *Remote Sensing for Environmental Monitoring, GIS Applications, and Geology IV*, Bellingham, WA.

Ehlers, M., S. Klonus, P. Johan Åstrand, and P. Rosso. 2010. Multi-sensor image fusion for pansharpening in remote sensing. *International Journal of Image and Data Fusion* 1 (1):25–45.

Evans, D. 2011. The internet of things: How the next evolution of the internet is changing everything. *CISCO White Paper* 1:14.

Gartner, I. *IT Glossary—Big Data*. Gartner, Inc. 2013 [cited 10 November 2015]. Available from http://www.gartner.com/it-glossary/big-data

Garzelli, A., L. Capobianco, L. Alparone, B. Aiazzi, S. Baronti, and M. Selva. 2010. Hyperspectral pansharpening based on modulation of pixel spectra. Paper read at *2nd Workshop on Hyperspectral Image and Signal Processing: Evolution in Remote Sensing (WHISPERS)*, Reykjavik, Iceland.

Ghaffar, M. A. A. and V. Tuong Thuy. 2015. Cloud computing providers for satellite image processing service: A comparative study. Paper read at *2015 International Conference on Space Science and Communication (IconSpace)*, Langkawi, Malaysia, August 10–12, 2015.

Gomez-Chova, L., D. Tuia, G. Moser, and G. Camps-Valls. 2015. Multimodal classification of remote sensing images: A review and future directions. *Proceedings of the IEEE* 103 (9):1560–1584.

Gubbi, J., R. Buyya, S. Marusic, and M. Palaniswami. 2013. Internet of Things (IoT): A vision, architectural elements, and future directions. *Future Generation Computer Systems* 29 (7):1645–1660.

Hampton, S. E., C. A. Strasser, J. J. Tewksbury et al. 2013. Big data and the future of ecology. *Frontiers in Ecology and the Environment* 11 (3):156–162.

Harris, J. R. and R. Murray. 1990. IHS transform for the integration of radar imagery with other remotely sensed data. *Photogrammetric Engineering and Remote Sensing* 56 (12):1631–1641.

Hashem, I. A. T., I. Yaqoob, N. B. Anuar, S. Mokhtar, A. Gani, and S. Ullah Khan. 2015. The rise of "big data" on cloud computing: Review and open research issues. *Information Systems* 47:98–115.

Helix Nebula. *The Science Cloud—Partnership between Big Science and Big Business in Europe* 2014 [cited 12 November 2015]. Available from http://www.helix-nebula.eu

Huadong, G. 2015. Digital Earth in the big data era. In *9th International Symposium on Digital Earth*, edited by ISDE. Halifax, Canada: ISDE.

Jansen, W. and T. Grance. 2011. Guidelines on security and privacy in public cloud computing. *NIST Special Publication* 800:144.

Kantardzic, M. 2011. *Data Mining: Concepts, Models, Methods, and Algorithms*: John Wiley & Sons.

Laben, C. A., V. Bernard, and W. Brower. 2000. Process for enhancing the spatial resolution of multispectral imagery using pan-sharpening. U.S. Patent. US & International: Laben et al., http://www.google.com/patents/US6011875

Leckie, D. 1990. Synergism of synthetic aperture radar and visible/infrared data for forest type discrimination. *PE&RS, Photogrammetric Engineering & Remote Sensing* 56 (9):1237–1246.

Lee, J.-G. and M. Kang. 2015. Geospatial big data: Challenges and opportunities. *Big Data Research* 2 (2):74–81.

Li, S., H. Wang, T. Xu, and G. Zhou. 2012. Application study on internet of things in environment protection field. In *Informatics in Control, Automation and Robotics*, edited by D. Yang. Berlin, Heidelberg: Springer.

Loncan, L., L. de Almeida, J. Bioucas-Dias et al. 2015. Hyperspectral pansharpening: A review. *IEEE Geoscience and Remote Sensing Magazine* 3 (3):27–46.

Ma, Y., H. Wu, L. Wang et al. 2015. Remote sensing big data computing: Challenges and opportunities. *Future Generation Computer Systems* 51:47–60.

Milenova, B. L. and M. M. Campos. 2005. Mining high-dimensional data for information fusion: A database-centric approach. Paper read at *2005 8th International Conference on Information Fusion*, Philadelphia, PA.

Mitsa, T. 2010. *Temporal Data Mining*, Chapman & Hall/CRC Data Mining and Knowledge Discovery Series. Boca Raton, FL: CRC Press.

Palsson, F., J. R. Sveinsson, M. O. Ulfarsson, and J. A. Benediktsson. 2014. Model based PCA/wavelet fusion of multispectral and hyperspectral images. Paper read at *2014 IEEE International Geoscience and Remote Sensing Symposium (IGARSS)*, Quebec City, Quebec, Canada, July 13–18, 2014.

Parsons, O. and G. A. Carpenter. 2003. ARTMAP neural networks for information fusion and data mining: Map production and target recognition methodologies. *Neural Networks* 16 (7):1075–1089.

Post, F. H., G. Nielson, and G.-P. Bonneau (Eds.). 2012. Data visualization: The state of the art. *Proceedings of the 4th Dagstuhl Seminar on Scientific Visualization*, Kluwer Academic Publishers, Boston.

Price, J. C. 1999. Combining multispectral data of differing spatial resolution. *IEEE Transactions on Geoscience and Remote Sensing* 37 (3):1199–1203.

Pugh, M., A. Waxman, and D. Fay. 2006. Assessment of Multi-Sensor Neural Image Fusion and Fused Data Mining for Land Cover Classification. Paper read at *2006 9th International Conference on Information Fusion*, Florence, Italy, July 10–13, 2006.

Ranchin, T. and L. Wald. 2000. Fusion of high spatial and spectral resolution images: The ARSIS concept and its implementation. *Photogrammetric Engineering & Remote Sensing* 66 (1):4–18.

Robila, S. A. 2006. Use of Remote Sensing Applications and Its Implications to the Society. In *IEEE International Symposium on Technology and Society, 2006. ISTAS 2006*, Flushing, NY.

Rokach, L. 2016. Decision forest: Twenty years of research. *Information Fusion* 27:111–125.

Shah, C. A., P. Watanachaturaporn, P. K. Varshney, and M. K. Arora. 2003. Some recent results on hyperspectral image classification. Paper read at *2003 IEEE Workshop on Advances in Techniques for Analysis of Remotely Sensed Data*, Greenbelt, MD, October 27–28, 2003.

Turner, V., J. F. Gantz, D. Reinsel, and S. Minton. 2014. *The Digital Universe of Opportunities: Rich Data and the Increasing Value of the Internet of Things*. http://www.emc.com/leadership/digital-universe/2014iview/index.htm.

Vermesan, O. and P. Friess. 2013. *Internet of Things: Converging Technologies for Smart Environments and Integrated Ecosystems*, River Publishers' Series in Information Science and Technology. Aalborg, Denmark: River Publishers.

Vongsingthong, S. and S. Smanchat. 2014. Internet of Things: A review of applications and technologies. *Suranaree Journal of Science & Technology* 21 (4):359–374.

Vrabel, J. 1996. Multispectral imagery band sharpening study. *Photogrammetric Engineering & Remote Sensing* 62 (9):1075–1083.

Wu, B. and S. Tang. 2015. Review of geometric fusion of remote sensing imagery and laser scanning data. *International Journal of Image and Data Fusion* 6 (2):97–114.

Yésou, H., Y. Besnus, and J. Rolet. 1993. Extraction of spectral information from Landsat TM data and merger with SPOT panchromatic imagery—A contribution to the study of geological structures. *ISPRS Journal of Photogrammetry and Remote Sensing* 48 (5):23–36.

Zhang, Y. 2008. Pan-sharpening for improved information extraction. In *Advances in Photogrammetry, Remote Sensing and Spatial Information Sciences: 2008 ISPRS Congress Book*, edited by Zhilin Li, Jun Chen, and Emmanuel Baltsavias. Boca Raton, FL: CRC Press, pp. 185–202.

Zhang, Y. and R. K. Mishra. 2014. From UNB PanSharp to Fuze Go—The success behind the pan-sharpening algorithm. *International Journal of Image and Data Fusion* 5 (1):39–53.

Zhou, Y. and F. Qiu. 2015. Fusion of high spatial resolution WorldView-2 imagery and LiDAR pseudo-waveform for object-based image analysis. *ISPRS Journal of Photogrammetry and Remote Sensing* 101:221–232.

Kluger, J. *Google* 2015 [cited 29 November 2015]. Available from http://world.time.com/timelapse/

Index

Printed and bound by CPI Group (UK) Ltd, Croydon, CR0 4YY

01/11/2024

01782617-0007